The Interpretation of Proton Magnetic Resonance Spectra

A Programmed Introduction

The Interpretation of Proton Magnetic Resonance Spectra

A Programmed Introduction

E. J. Haws *The Polytechnic, Wolverhampton*
R. R. Hill *The Open University, Bletchley, Buckinghamshire*
D. J. Mowthorpe *The Polytechnic, Sheffield*

London · New York · Rheine

Heyden & Son Ltd., Spectrum House, Alderton Crescent, London NW4 3XX.
Heyden & Son Inc., 225 Park Avenue, New York, N.Y. 10017, U.S.A.
Heyden & Son GmbH, 4440 Rheine/Westf., Münsterstrasse 22, Germany.

© Heyden & Son Ltd., 1973

All Rights Reserved. No part of this publication may be reproduced, stored in a retrieval system, or transmitted, in any form or by any means electronic, mechanical, photocopying, recording or otherwise, without the prior permission of Heyden & Son Ltd.

ISBN 0 85501 063 0

Printed in Great Britain by J. W. Arrowsmith Ltd., Bristol BS3 2NT

Contents

Preface vii
How to Use the Programme xi
Guide to Programme Timing xii

Part 1: BASIC PRINCIPLES
Aim, New Terms and Prior Knowledge 3
Objectives 4
Introduction 5
Multiple Choice Test 5
Programme 7
Revision Summary 14
Questions 15
Further Reading 15

Part 2: CHEMICAL SHIFT AND INTEGRATION
Aim, New Terms and Prior Knowledge 19
Objectives 20
Introduction 21
Multiple Choice Test 22
Diamagnetic Shielding 25
Integration 30
Chemical Shift and its Units 34
The Empirical Use of Chemical Shift and Integration . . 42
Shielding and Deshielding Effects of Multiple Bonds . . 48
Revision Summary 55
Questions 58
Further Reading 60

Part 3: SPIN–SPIN COUPLING
Aim, New Terms and Prior Knowledge 65
Objectives 66
Introduction 67
Multiple Choice Test 67
The Origin of Spin–Spin Coupling 70
The Branching Method 76
NH and OH Protons 81
Examples 85
Coupling Constants 90
Examples 93
Revision Summary 97
Questions 98
Further Reading 100

Part 4: COMPLEX (SECOND-ORDER) SPECTRA
    Aim, New Terms and Prior Knowledge 103
    Objectives 104
    Introduction 105
    Multiple Choice Test 105
    Complex (Second-Order) Spectra 107
    Aids to the Interpretation of Complex Spectra 113
    Revision Summary 126
    Questions 127
    Further Reading 127

Part 5: TEMPERATURE DEPENDENT SPECTRA AND GEMINAL NON-EQUIVALENCE
    Aim, New Terms and Prior Knowledge 131
    Objectives 132
    Introduction 133
    Multiple Choice Test 133
    Temperature Dependent Spectra 137
    Geminal Non-Equivalence 142
    Revision Summary 153
    Questions 154
    Further Reading 155

Part 6: PROBLEMS
    Objective 159
    Introduction 160
    Problems 161
    Further Reading 173

Answers to Multiple Choice Tests 175

Selected Answers to Questions 176

List of Compounds 177

Subject Index 179

CHARTS 1 AND 2 . . . . *facing p.* 180

Preface

Nuclear magnetic resonance (n.m.r.) spectroscopy, particularly that of protons, is one of the most useful techniques for the elucidation of chemical structure and reaction mechanisms. Though finding greatest application in a research laboratory, it now forms an important part of most undergraduate curricula. This is because it contributes significantly to the evidence on which many chemical structures, theories and concepts are based. Thus, like other experimental data, it helps the student to view accepted theories of the subject more critically. But here lies a dilemma. Few practising chemists would deny that although the principles of a spectroscopic method can often be mastered in a short time, real competence at evaluating data only comes with experience of its use, preferably in one's own work. Furthermore, such experience is accumulated at a rate which depends on the frequency with which the particular method is used. So by the time the undergraduate masters a technique like n.m.r. sufficiently well to be able to assess the data critically he may well have already completed most of his course. Most books on interpretive n.m.r. spectroscopy leave this problem firmly in the hands of teachers, many of whom no doubt go some way to solving it by the incorporation of simple exercises, practical work and tutorials. Such facilities are not normally available to the isolated learner, such as for example a research worker who finds he has to use the technique and who has to draw on his experience as an undergraduate, or a non-graduate assistant who meets n.m.r. for the first time in the laboratory.

The general aim of this book is to combine effective teaching of the subject matter with a means of rapid accumulation of experience in interpreting spectra. By using a linear programmed style we are able to supply a continuous feedback to the student on his understanding of the material and also provide step-by-step guidance in the interpretation of spectra.

The book is divided into six parts, each a self-contained teaching 'package' assuming only a knowledge of what is contained in previous parts. It starts with a non-mathematical introduction to basic principles where many magnetic nuclei are mentioned but attention is quickly focused on the proton. This is followed in Part 2 by chemical shift and integration. Part 3 covers first-order spin—spin coupling and, in both this and the previous part, the importance of the correct use of empirical data is emphasized. Part 4 extends coupling to second-order and also includes a small section on methods of extracting information from complex spectra, such as spin-decoupling and the use of deuterium. The next part considers two aspects of the subject which are no longer 'sophisticated' applications, but feature prominently in any area where n.m.r. is used: namely, temperature dependent spectra and the special effects arising from chirality. The final part is solely concerned with the interpretation of spectra and consists of worked examples. In addition to the straight programmed text, Parts 1 to 5 also include the following:

(i) a statement of aims, new terms and concepts and assumed prior knowledge;
(ii) a list of objectives;

(iii) an introduction which forms a link with the previous part (except in the case of Part 1);
(iv) a self-assessment test;
(v) remarks on pre-test scores;
(vi) a summary;
(vii) a self-assessment test;
(viii) remarks on post-test scores;
(ix) a bibliography;
(x) problems;
(xi) a link paragraph to the next part.

The correlation charts are at the end of the book and inserted so that they can be pulled out to be permanently on display whilst working through the programme.

We trust this text will be of value to a wide range of chemists and chemistry students. The overall aim of the book is to enable the reader to become competent in the interpretation of nuclear magnetic resonance spectra and to be able to identify situations where n.m.r. could be used with advantage. It is particularly directed to the student and we have attempted to adapt it for classroom use. Parts 1, 2 and 3 serve as an introduction to the subject whereas the remaining parts are more suitable for use at a later stage in the curriculum. Those already familiar with nuclear magnetic resonance but who wish to extend their knowledge in the direction of the latter part of the programme, may find Parts 1, 2 and 3 useful revision material prior to studying Parts 4, 5 and 6.

In the classroom, the programme can be used as a component in a course in n.m.r. spectroscopy. The main facts and concepts can be left to the programme, freeing the teacher to concentrate on applications in tutorials or cope with individual students' queries. The programme allows for different rates of progress; slow students can be assisted by the teacher and, if necessary, encouraged to catch up with the average worker in their own time. Students who make rapid progress can be referred to the questions and the bibliography at the end of each part. The programme can also be used for revision purposes or as an aid to the student who has missed several weeks' work.

We also hope this book will be of considerable help to those who cannot conveniently benefit from classroom/laboratory instruction and have to study or catch up in isolation. Such would be the case of an individual who finds that he has to use n.m.r. data or is unsure whether it can be used with advantage for his particular problem. The book could be used as an up-dating or revision exercise, or perhaps as an introduction to the subject for the first time. It is largely self-teaching for both kinds of use but readers need to have some knowledge of basic chemistry before starting Part 1.

Programmed Learning
Although interpretive spectroscopy lends itself to programmed learning this is only part of the reason why we chose to use it in our book. It is now becoming widely recognized that the teaching effectiveness of written material can be significantly improved if it is made student-active rather than passive, and if the student can obtain feed-back while he is working, and thereby continuously assess his progress.

Programmed learning encourages this to happen. The subject matter is divided up into units known as frames. At the end of each frame the student finds out immediately whether he has understood the point being made. If he has not, he can rectify matters there and then by re-reading the material more carefully and taking note of any remedial comments. Nevertheless, though it is more self-teaching than most conventional books, a linear programme can never be regarded as a satisfactory medium on its own. This is because it normally allows little room for the student to develop his own thoughts on the subject matter. It is a means of rapidly acquiring a basic knowledge and understanding of a topic which can then be applied in the less intellectually restricted atmosphere of tutorial work and discussion with one's colleagues or fellow students.

This programme is different in many respects from earlier published programmes. We have in the main avoided very short frames, that is to say, frames of low intellectual content. The type of response is highly varied and we have avoided completely the 'missing word' type. We have found that one consequence of using linear programmed style (which guides the students' thoughts step by step through conceptually difficult areas of the topic) is that it has been necessary to introduce generalities at the expense of rigour in one or two places. We felt this was justified where a lengthy digression for the sake of precision would be an undesirable interruption in the main argument.

Finally, the main characteristic of programmed material is that it has been tested and proved to be effective. This book has undergone an objective validation programme and has been tested and revised three times. The evaluation was carried out with the cooperation of over 100 college and industry-based students involving widely different academic backgrounds. Students used the criterion tests associated with Parts 1 to 5 to assess their ability to achieve the objectives of each part before and after working through the relevant material. The final version of the programme enabled the average student with little or no knowledge of the subject to increase his test performance as follows:

	Pre-test score (%)	*Post-test score (%)*
Part 1	35	95
Part 2	10	85
Part 3	20	80
Part 4	25	85
Part 5	30	80

Acknowledgements

The validation was only possible through the voluntary help of many students and their supervisors in the 30 institutions which participated. We gratefully acknowledge the considerable time and effort they spent on the manuscript and the constructive comments they offered for its improvement. Drs. R. W. Alder and T. Threlfall made particularly detailed comments and valuable suggestions, as did many of our colleagues at Wolverhampton Polytechnic.

June, 1973

E. J. Haws
R. R. Hill
D. J. Mowthorpe

How to Use the Programme

You will probably find the best approach to each part is as follows. Take note of assumed prior knowledge, read the introduction and attempt the self-assessment test. The test is important because it gives you a more objective view of your prior knowledge and, by comparing it with the score you obtain after working through the subsequent programmed material, you will have direct evidence of what has been achieved. The programme itself is written in a fairly common format and we suggest you read it in the following way:

(i) Cover the page with a sheet of paper and pull it down to the first line drawn across the page.
(ii) Read the material exposed and attempt the question (or in some cases, choose the most appropriate word from those printed in italic type), writing your answer on a separate sheet of paper.
(iii) Pull the covering sheet down to the next full line, revealing the next frame and the answer to the first.
(iv) Check your answer to frame 1 with that given.
   (a) If it is correct read the material of frame 2 and continue as from (ii) above.
   (b) If it is incorrect take note of any remedial comment given with the answer or re-read frame 1 and try to understand the reason for your incorrect response. Then continue as from (ii) above.

When you have worked through the programmed material in a part, read the summary and attempt the self-assessment test, marking your answers as before.

The continuous activity in programmed instruction is necessarily more demanding than reading a normal text and you should not study for more than about *40 minutes* without some kind of a break. Each part of the programme has been divided into sections for this purpose (except Part 1 which is short) though you may find you can cover more than one section at a sitting. On the other hand the overall aim of the programme will not be achieved if you spread the work on either of the two halves (Parts 1, 2 and 3 OR Parts 4, 5 and 6) over too long a period. You should aim to complete each of these halves within a period of less than three months. The average times spent on each section by students who validated the programme are given on p. xii. We hope these figures will be helpful in planning your work.

However the programme is used, the retention of the subject matter will depend very much on how frequently you use n.m.r. spectroscopy afterwards. As with proficiency in languages, continuous practice is necessary not only for improvement but for maintaining a level of competence. In this regard it is perhaps worthwhile noting that much intellectual satisfaction can be had from practising one's skill on problems in spectroscopy and, where a lull in application of the technique arises, the solving of such puzzles as are presented in books given in the bibliography for Part 6 can be not only a useful activity but an enjoyable one as well!

Guide to Programme Timing

Here as a guide are the times in minutes spent on average by the students who helped to validate the book.

Part 1
    Programme 40

Part 2
    Diamagnetic Shielding 20
    Integration 25
    Chemical Shift and its Units 40
    The Empirical Use of Chemical Shift
        and Integration 35
    Shielding and Deshielding Effects of
        Multiple Bonds 35

Part 3
    The Origin of Spin–Spin Coupling . . 25
    The Branching Method 25
    NH and OH Protons 15
    Examples 35
    Coupling Constants 20
    Examples 35

Part 4
    Complex (Second-Order) Spectra . . 35
    Aids to the Interpretation of Complex
        Spectra 50

Part 5
    Temperature-Dependent Spectra . . . 25
    Geminal Non-Equivalence 45

Part 6
    Problems 90

PART 1
BASIC PRINCIPLES

Aim
This section should give the student the minimum knowledge of the basic principles of nuclear magnetic resonance considered necessary to understand how the features of first-order spectra arise (dealt with in later parts). The section leads him up to the point where he appreciates how an absorption band arises when a sample containing protons is irradiated with electromagnetic radiation of constant frequency in a varying magnetic field. He will become aware of the scope of the method from a theoretical point of view and some of its practical limitations.

New Terms and Concepts
Orientation of spinning atomic nuclei in a magnetic field.
Absorption of electromagnetic radiation leading to promotion of a nucleus from a ground to an excited state.

Relaxation, saturation, hertz, megahertz, absorption band.

Prior Knowledge
Elementary atomic structure and the meaning of the term 'isotope'.

Objectives for Part 1
When you have completed this part you should be able to:
1. Identify the feature of atomic or molecular structure which gives rise to n.m.r. absorption.
2. Given the spin properties of an atomic nucleus:
   (a) predict whether it could undergo nuclear magnetic resonance;
   (b) predict the number of orientations it could adopt in a magnetic field.
3. State the relationship between frequency of radiation absorbed and the applied magnetic field for nuclear magnetic resonance and indicate how they are varied in the running of a routine spectrum.
4. Recognize correct statements about suitable experimental conditions for the measurement of a routine spectrum of an organic substance.
5. Select the correct region of the electromagnetic spectrum suitable as a radiation source for n.m.r. spectroscopy.
6. State the effect of absorption of radiation on the orientation of protons in a magnetic field.
7. Distinguish between the spectra of samples measured in the solid state and in solution.

Introduction

The purpose of the first section of the programme is to describe the essential basis of nuclear magnetic resonance (n.m.r.) in simple terms, and only to an extent considered necessary to enable you to acquire an understanding of the material to follow,* which will deal more directly with the interpretation of spectra.

At the end of the section, you will be able to recognize the features which n.m.r. has in common with other forms of spectroscopy. You will also be aware of the types of compounds which can be examined, and some of the important practical limitations of the method.

Multiple Choice Test

Before starting the programme itself, try the following multiple choice test. Write your answer in a vertical column on a separate sheet of paper which you have prepared as shown here. Marking instructions will follow the test.

Answer	x Correct +1	y Incorrect $-\frac{1}{2}$

1. Nuclear magnetic resonance is a phenomenon connected with the interaction of a nucleus with a magnetic field. All nuclei are capable of resonance **except** those with:
   (a) an even mass number and even atomic number;
   (b) an even mass number and odd atomic number;
   (c) an even atomic number and odd mass number.

2. Which of the following statements is **false**?
   (a) The n.m.r. phenomenon is limited to one nucleus only, namely the proton.
   (b) The frequency required to bring a proton into 'resonance' is directly proportional to the applied field.
   (c) N.m.r. spectra are usually obtained from solutions or pure liquids.

3. The magnetic moment vector of the proton can assume:
   (a) 1, (b) 2, (c) 3 orientations in a magnetic field.

4. The relevant expression describing the energy difference between orientations of a proton in a magnetic field is:

   $\Delta E =$ (a) $\dfrac{2\mu H_0}{h}$ (b) $\dfrac{2H_0}{h\mu}$ (c) $2\mu H_0$

   where μ = the magnetic moment, h = Planck's constant and H_0 = applied field.

5. The magnetic fields of commercial n.m.r. instruments are of such magnitude that absorption of energy in the (a) visible, (b) microwave, (c) radio frequency range of the electromagnetic spectrum is necessary to attain the resonance condition.

6. Absorption of energy causes the number of protons with their magnetic moment vectors aligned:
   (a) with the field to increase;
   (b) against the field to increase;
   (c) against the field to decrease.

---

* A bibliography at the end of Part 1 refers to more detailed accounts of theoretical aspects of n.m.r.

7. Which of the following statements is **true**?
   (a) Samples in the liquid state or in solution give rise to broader n.m.r. absorption bands than samples in the solid state.
   (b) Samples in the liquid state or in solution give rise to weaker n.m.r. absorption bands than samples in the solid state.
   (c) Samples in the solid state give rise to broader n.m.r. absorption bands than samples in the liquid state or in solution.
8. In practice an n.m.r. absorption spectrum is **usually** obtained by (a) varying both field and frequency; (b) varying the field, keeping frequency constant; (c) varying the frequency, keeping the field constant.
9. Which of the three solvents given below is most commonly used when measuring the n.m.r. spectra of protons?
   (a) $CDCl_3$ (b) $CH_3-CO-CH_3$ (c) C_2H_5OH
10. In order to measure an n.m.r. spectrum of a substance one usually needs
    (a) 0.005–0.025 mg, (b) 5–25 mg, (c) 5–25 g of the material.

Comments on marks

Compare your answer column with the vertical sequence of answers given for this test on p. 175. Insert one mark in column x for every correct and ½ mark in y for every incorrect answer. Unanswered questions score zero. Subtract the total of column y from that of column x, and compare the result with the following remarks. Do not seek the correct answers to incorrectly answered questions unless your mark falls into the third category.

Less than 3: Proceed with the programme.

3–6½: Although you have some knowledge and understanding of n.m.r. already it is not sufficient to guarantee a full grasp of what is to come in subsequent parts. Work through this part – it shouldn't take you long.

7 and over: Your knowledge of basic principles is adequate for you to continue directly with Part 2. However, you should check up on those points which led you to a wrong answer in the test.

PART 1: BASIC PRINCIPLES

F1 The atomic structure of a particular element can be pictured as a nucleus of protons (positively charged) and neutrons (uncharged) surrounded by negatively charged electrons. The nucleus is therefore *positively/negatively* charged.

F2 An element can exist as any one of a number of isotopes depending on the number of neutrons present in the nucleus. Nuclear magnetic resonance, as its name implies, results from a property of atomic nuclei.

Is the following statement true or false? In discussing nuclear magnetic resonance (n.m.r.) we shall be concerned primarily with particular isotopes rather than particular elements.

    A1 positively.

F3 However, by no means all atomic nuclei can undergo nuclear magnetic resonance. It turns out that isotopes with either an odd atomic number (odd number of protons in the nucleus) or odd mass number (odd total number of protons and neutrons) for example ^{1}H, ^{19}F, ^{31}P, ^{14}N, show the phenomenon, whereas those with even atomic number and mass number, for example ^{12}C, ^{16}O, ^{32}S do not.

Is it correct to say that all nuclei with odd numbers of neutrons can undergo nuclear magnetic resonance?

    A2 True.

F4 Give two examples of nuclei which can undergo n.m.r. and two examples of nuclei which cannot.

    A3 Yes. An odd number of neutrons can be associated with an odd number of protons, in which case the nucleus has an odd atomic number; or it can be associated with an even number of protons, in which case it has an odd mass number.

F5 Nuclei which are capable of resonance behave like spinning charged particles (see Fig. 1.1). Basic laws of physics tell us that such a particle behaves like a small magnet generating its own magnetic field and that consequently it will interact with any other magnetic field.

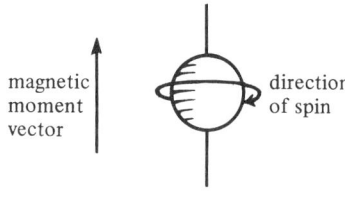

Fig. 1.1.

Which of the following nuclei will interact with an applied magnetic field?
    ^{1}H, ^{12}C, ^{16}O, ^{19}F

A4 Those which can (n.m.r. active): two from ^1H, ^{19}F, ^{31}P, ^{14}N.
Those which cannot (n.m.r. inactive): two from ^{12}C, ^{16}O, ^{32}S.

F6 A normal bar magnet may be placed in any one of an infinite number of orientations relative to an external magnetic field. Each position of the magnet in this field will have a different potential energy ranging from the lowest when the magnet is aligned with the field, to the highest when aligned against the field. The nuclear magnet, on the other hand, is restricted by quantum considerations to a limited number of orientations in an external field.

Which can have the greatest number of potential energy values in an external magnetic field: (a) a bar magnet, (b) a nuclear magnet?

A5 ^1H, ^{19}F.

F7 The number of orientations which a nucleus can take up in a magnetic field is determined by a nuclear property known as the spin quantum number (I). The spin quantum number for any particular nucleus can have one of the values 0, 1/2, 1, 3/2, 2, 5/2 etc., and the number of possible orientations is $2I + 1$. (Nuclei with $I = 0$ are those which are n.m.r. inactive.)

How many orientations may be taken up by nuclei which have spin quantum numbers equal to:
 (a) 1/2, (b) 3/2?

A6 (a) A bar magnet.

F8 How many orientations can the ^{14}N nucleus (spin quantum number of 1) assume in an applied magnetic field?

A7 (a) 2, (b) 4.

F9 Most nuclear magnetic resonance data amassed so far have been concerned with protons, which occur in all organic compounds and have simple magnetic properties. Accordingly we shall concentrate on this nucleus although nuclear magnetic resonance is now an important technique for many other nuclei.*

The spin quantum number for the proton is 1/2. How many orientations can it assume in an applied magnetic field?

---

* In practice 'n.m.r.' has been almost synonymous with proton magnetic resonance as far as the organic chemist is concerned. Current developments, however, particularly in instrument design, now hold the promise of ^{13}C magnetic resonance spectra becoming widely available in the foreseeable future (in spite of the fact that only one carbon atom in a hundred has a ^{13}C nucleus). It has been shown already that ^{13}C spectra are extremely informative and the developments in this area are likely to constitute a major advance in structure-determining techniques (G. C. Levy and G. L. Nelson, *$^{13}C NMR$ for Organic Chemists*, John Wiley & Sons, New York, 1972). It is important, therefore, to bear in mind that throughout this programme the use of the term 'n.m.r.' is limited strictly to proton magnetic resonance.

A8 Three.

F10 A proton, then, can be considered as a small magnet which can assume two orientations in a magnetic field. In one of these orientations its magnetic moment vector is aligned **with** and in the other aligned **against** the applied field.* No other orientations are possible. Of the two, the former is the more stable. Therefore in a magnetic field a proton can be in one of two energy levels corresponding to 'alignment' with or against the field. Does the higher energy level correspond to alignment with or against the applied field?

A9 Two. The number of orientations $= 2I + 1$.

F11 We can represent this on an energy diagram (Fig. 1.2) where ΔE is the energy difference between the two levels. This quantity is directly proportional to the strength of the applied field, H_0, and is given by the expression:

$$\Delta E = 2\mu H_0 \qquad (1)$$

where μ, the magnetic moment of the nuclear magnet, is a constant for a particular nucleus.

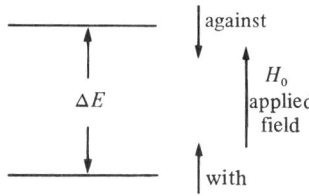

Fig. 1.2.

Thus the greater the applied field the *greater/smaller* the energy gap between the two possible states of the proton.

A10 Against. The frame told you that alignment with the field is the more stable of the two possible positions. It follows that alignment against the field is the least stable, and corresponds to the higher energy level.

F12 A proton in the lower energy level can be promoted to the upper level by supplying the exact quantity of energy ΔE. Conversely, if we have a source of energy ΔE available to the proton in a magnetic field H_0, will it be absorbed if:

$$\Delta E = \text{(a)} \ \frac{2H_0}{\mu}, \ \text{(b)} \ \frac{2\mu}{H_0}, \ \text{or (c)} \ 2\mu H_0?$$

A11 greater. E is directly proportional to the applied field.

F13 Electromagnetic radiation (of which visible light is an obvious but very small part) is a form of energy whose magnitude, E, is given by the expression:

$$E = h\nu \qquad (2)$$

where h is Planck's constant and ν is the frequency of radiation.

* Strictly speaking, 'almost with' and 'almost against'.

Hence if we irradiate a proton in its lower energy level in a magnetic field with electromagnetic radiation it will be excited to a higher level provided that ΔE (the difference between the energy levels) $= h\nu$.

Express this condition in terms of ν (the frequency), H_0 (the applied field), μ (the magnetic moment) and h (Planck's constant); i.e. complete the expression:

$$\nu = ? \text{ in terms of the above parameters.}$$

A12 (c) $\Delta E = 2\mu H_0$.

F14 Nuclear magnetic resonance is only relevant if we are dealing with substances containing nuclei whose spin quantum numbers are *greater than/equal to* zero. Which of the commonly occurring nuclei ^{12}C, ^{1}H, ^{16}O, ^{14}N are capable of showing nuclear magnetic resonance?

A13
$$\nu = \frac{2\mu H_0}{h} \tag{3}$$

F15 The frequency required for absorption of energy (i.e. for resonance) for a proton immersed in a field of 14 000 G (which is typical of that produced by powerful commercial magnets used in n.m.r. spectrometers) lies in the range normally used by radio and television at 60 MHz. This comes from Eqn. (3). Other more expensive instruments use fields in the region of 23 000 G. Using the latter machines, the frequency necessary to observe proton resonance is *40/100* MHz.

A14 greater than; ^{1}H, ^{14}N.

F16 Let us digress from protons for a moment. The values quoted in frame 15 were derived from Eqn. (3) $[\nu = 2\mu H_0/h]$. You recall that μ is a fundamental nuclear constant. It turns out that μ for ^{31}P is less than half of that for ^{1}H. Consequently a spectrometer operating at 14 000 G (and therefore using a 60 MHz frequency for detecting protons) should operate at a frequency *less than/greater than* 30 MHz for ^{31}P.

A15 100 MHz. Equation (3) indicates that the frequency for proton resonance is directly proportional to the applied field. Therefore the frequency for proton resonance in a field of 23 000 G must be greater than that in a field of 14 000 G.

F17 If it were possible to use visible light, which has a **very much** shorter wavelength and consequently **very much** higher frequency than that usually employed in n.m.r. experiments, one would need an *exceedingly powerful/very weak* magnetic field to obtain the resonance condition.

A16 less than. Again this comes from Eqn. (3).

F18 A proton immersed in a field of 14 000 G can be 'flipped' from a state where its magnetic moment vector is aligned *with/against* the field to one where it is aligned *with/against* the field by the absorption of electromagnetic radiation in the *visible/radio frequency* of the spectrum.

A17 exceedingly powerful. (It is impossible to make magnets capable of generating anything like the required field strength.)

F19 Before we irradiate a sample of a compound containing protons in a magnetic field, the protons will have populated the two states of slightly different energy according to thermodynamic principles (Boltzmann distribution) and there will be a slight excess in the lower energy level. If we now irradiate the sample, energy $\Delta E = h\nu$ will be absorbed and the population of the upper state will *increase/decrease*.

A18 with; against; radio frequency.

F20 Eventually, the original small excess of nuclei in the ground state is eliminated and net absorption ceases, **unless** there is some mechanism to allow nuclei to return to the ground state. Fortunately, there is a means of maintaining the ground state excess, whereby protons in the excited state release energy ΔE to the environment. This process is known as RELAXATION. If, under some circumstances this mechanism cannot cope, there is no net absorption and the sample is said to be SATURATED.

State whether the following statements are true or false.
(a) When a sample containing protons is immersed in a magnetic field more protons are aligned against the field than with the field.
(b) Absorption of energy increases the number of protons aligned against the field.
(c) Saturation occurs when the upper state is drastically depopulated.

A19 increase. Absorption of energy leads to the promotion of protons from the lower to higher energy level.

F21 If we plot absorption against frequency of radiation we will obtain an absorption curve as shown in Fig. 1.3. The width of the absorption band depends on the relaxation process. In the case of liquid samples or solutions the bands are of convenient widths for analysis. Samples in the solid state, however, relax through a different mechanism which results in very broad lines.

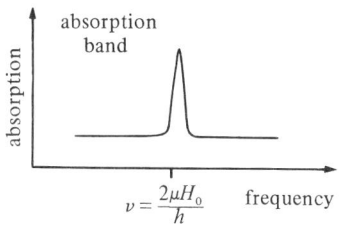

Fig. 1.3. Sample in a field H_0.

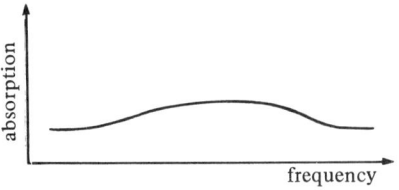

Fig. 1.4. Typical absorption of a sample in the *solid/liquid* state.

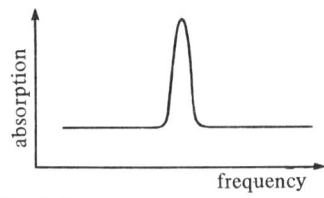

Fig. 1.5. Typical absorption of a sample in the *solid/liquid* state.

A20 (a) False; the opposite is true, see frame 19.
(b) True.
(c) False; saturation occurs when relaxation is not sufficiently effective to maintain an excess of protons in the lower energy state, the two states become equally populated and net absorption reduces to zero.

F22 You recall that the resonance condition is represented by the expression:

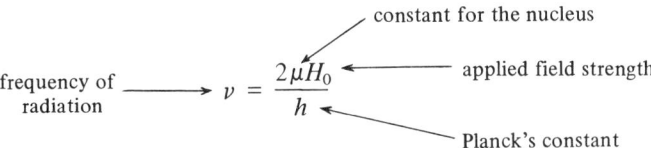

Thus the frequency necessary for absorption is directly proportional to the applied field. Because of this direct proportionality we could equally well irradiate the sample with radiation of constant frequency v, and vary the field until we reach the value $H_0 = ?$, where absorption would occur. This is what is usually done in practice.

A21 Fig. 1.4 – solid; Fig. 1.5 – liquid.

F23 The sample is placed in a magnetic field and irradiated at frequency v. The field is varied over a small range, and an electronic device detects where, on a field scale, absorption of the radiation occurs. This is displayed on a chart as an absorption band. Look at Fig. 1.3 in frame 21 and complete Fig. 1.6 showing the absorption at constant frequency.

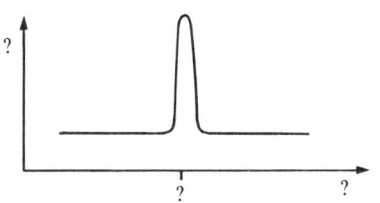

Fig. 1.6. Sample irradiated with frequency v.

A22 $$H_0 = \frac{hv}{2\mu}$$

F24 In order to measure the proton magnetic resonance absorption of a compound, the sample, preferably in the *liquid/solid* state or in solution, is immersed in a magnetic field and irradiated with radio frequency radiation of constant frequency. The field is then varied, absorption occurs at a particular field value and is displayed as an absorption curve on a chart which constitutes a plot of absorption versus *field/frequency*.

A23

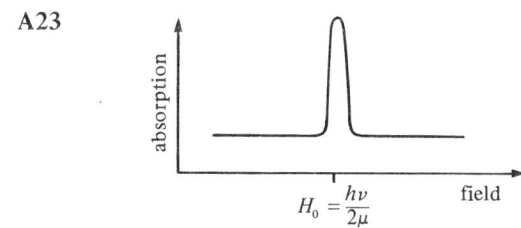

F25 Suppose one had a solid and wished to measure its proton magnetic resonance spectrum. First, the sample must be dissolved in a suitable solvent. It is generally preferable to use a solvent which does not have a **proton** n.m.r. spectrum. On these grounds, which of the following common solvents would be preferred? Carbon tetrachloride, chloroform, tetrachloroethylene, benzene, water, carbon disulphide.

A24 liquid; field.

F26 The number of common solvents used in n.m.r. is therefore limited. However, deuterated solvents, i.e. solvents containing hydrogen as its deuterium isotope, are now available. (Although deuterium has a magnetic moment, it does not absorb in the same region as a proton.)

Substitution of hydrogen by deuterium does not appear to affect solvation properties, and useful deuterated derivatives of those solvents you rejected in frame 25 are available. Write out the formulae of these deuterated derivatives.

A25 Carbon tetrachloride, tetrachloroethylene, carbon disulphide; the other three contain protons.

F27 Having chosen a suitable solvent, a solution is made up with the strongest possible concentration, ideally greater than 5%. This is poured into a tube sealed at one end and placed between the poles of the magnet as shown in Fig. 1.7.

Suppose a solid was 1% soluble in carbon tetrachloride and 10% soluble in chloroform. Suggest a solvent suitable for n.m.r.

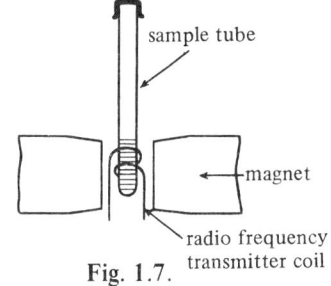

Fig. 1.7.

A26 $CDCl_3$, D_2O, C_6D_6.

F28 Approximately $1/3$ cm^3 of such a solution is required for a measurement. What is the approximate weight of a sample in mg, that would be required to obtain its n.m.r. spectrum?

A27 CDCl$_3$. Solubility in a deuterated solvent would be the same as in its proton analogue.

F29 Once the tube containing the solution is between the poles it is irradiated with radio frequency radiation, the magnetic field is varied and an absorption spectrum is recorded on the chart.

A28 10 to 20 mg. Frame 28 implies the minimum concentration should be about 5%. 1/3 cm^3 of such a solution therefore contains 1/3 × 50 mg = 16.6 mg. This figure is unrealistically precise in the present context.

You may find it useful at this point to read through the following revision summary.

Revision Summary

Nuclear magnetic resonance is a phenomenon limited to compounds containing isotopes with either odd atomic number or odd mass number. Examples include ^{1}H, ^{19}F, ^{31}P, ^{14}N. (^{12}C, ^{16}O and ^{32}S have no spin).

Nuclei which are capable of resonance behave as though they were small magnets which, when placed in an external magnetic field can assume only $2I + 1$ orientations with respect to the direction of the field, where I is the spin quantum number of the nucleus. The proton, ^{1}H, which has a spin quantum number $I = \frac{1}{2}$, can assume two orientations in a magnetic field. These correspond to two energy states, alignment with the field being the lower and alignment against the field the higher. The energy difference between the states is related to the magnitude of the external field H_0 (see Fig. 1.8).

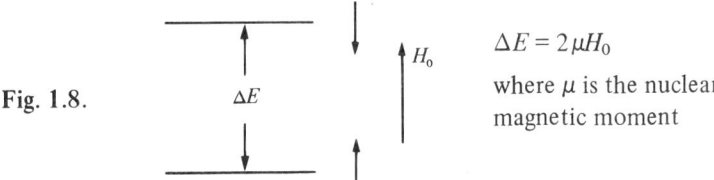

Fig. 1.8. ΔE H_0 $\Delta E = 2\mu H_0$
where μ is the nuclear magnetic moment

Transition from the lower to the higher energy state can be promoted by the absorption of electromagnetic radiation of frequency ν by the nucleus when:

$$\Delta E = h\nu \quad (h \text{ being Planck's constant})$$

Thus $\nu = 2\mu H_0/h$

and $\nu \propto H_0$

The magnetic fields employed in commercial instruments are of such magnitude that radiation in the radio frequency range is required to attain the resonance condition.

Relaxation is the process by which nuclei in the excited state can return to the ground state. Absence of relaxation leads to saturation, a situation where there is no net excess of nuclei in the ground state and net absorption ceases.

The form of relaxation is critical to the appearance of an absorption band. Solid samples give broad diffuse bands whereas solutions and liquids give sharp narrow bands.

An n.m.r. absorption band is usually recorded by sweeping the magnetic field whilst irradiating the sample with radiation of constant frequency. Absorption of the radiation occurs when the field reaches the value $h\nu/2\mu$.

The most common procedure for measuring n.m.r. spectra involves dissolving the sample in a suitable solvent. Solvents containing protons are avoided. Carbon tetrachloride and deuterochloroform are used most frequently.

Now have another go at the multiple choice test at the beginning of this part. Write your answers in a vertical column and mark the script as before. Then consult the remarks below.

0–6: Consult your instructor as to whether you should go on or seek extra tuition on basic principles of the subject. If this is inappropriate, reread the frames pertinent to those questions which you answered incorrectly and, when you are satisfied you fully understand the material, read the paragraph at the end of the part and start Part 2.

More than 6: Revise the material pertinent to the questions you answered incorrectly. If your instructor does not want you to attempt the questions which follow or consult the bibliography, read the paragraph at the end of this part and continue with Part 2.

Questions

1. Derive the expression $H_0 = h\nu/2\mu$ for the field at which nuclear magnetic resonance may occur.

2. State for each of the following compounds, the number of atoms which could give rise to a nuclear magnetic resonance signal. (Consider the commonest isotope in each case.)

    (a) CH_3COCH_3 (b) [imidazole-type structure with H, H on C=C, N, N–H, C–CO$_2$H]   (c) FSO_3H

3. Illustrate diagrammatically the origin of an absorption band in nuclear magnetic resonance, when the sample is subject to a varying magnetic field whilst under radiation by electromagnetic radiation of constant frequency.

4. What happens to an n.m.r. absorption band when the sample is 'saturated'? Explain.

5. Comment on the need for a solvent and the type of solvents used in n.m.r.

Further Reading

L. M. Jackman and S. Sternhell, *Applications of Nuclear Magnetic Resonance Spectroscopy in Organic Chemistry,* Pergamon Press, Oxford, 1969, Chapters 1.1 and 1.2.
D. W. Mathieson, *Nuclear Magnetic Resonance for Organic Chemists,* Academic Press, New York, 1967, Chapter 1.

J. W. Emsley, J. Feeney and L. H. Sutcliffe, *High Resolution Nuclear Magnetic Resonance Spectroscopy*, Volume 1, Pergamon Press, Oxford, 1965, Chapters 1 and 2.

W. W. Paudler, *Nuclear Magnetic Resonance*, Allyn and Bacon, Boston, 1971, Chapter 1.

Introductory sections of the following:

H. G. Hecht, *Magnetic Resonance Spectroscopy*, John Wiley & Sons, New York, 1967.

A. Carrington and A. McLachan, *Introduction to Magnetic Resonance*, Harper & Row, New York, 1967.

D. Chapman and P. D. Magnus, *Introduction to Practical High Resolution Spectroscopy*, Academic Press, London, 1966.

J. R. Dyer, *Application of Absorption Spectroscopy of Organic Compounds*, Prentice-Hall, New Jersey, 1965, Chapter 4.

R. M. Silverstein and G. C. Bassler, *Spectrometric Identification of Organic Compounds*, John Wiley & Sons, New York, 1967, Chapter 4.

J. C. P. Schwarz, *Physical Methods in Organic Chemistry*, Oliver and Boyd, Edinburgh, 1964, Chapter 5.

D. H. Williams and I. Fleming, *Spectroscopic Methods in Organic Chemistry*, McGraw-Hill, London, 1966, Chapter 4.

D. J. Pasto and C. R. Johnston, *Organic Structure Determination*, Prentice-Hall, New Jersey, 1969, Chapter 5.

S. F. Dyke, A. J. Floyd, M. Sainsbury and R. Theobald, *Organic Spectroscopy – An Introduction*, Penguin, 1971, Chapter 4.

You should now be in a position to appreciate that certain nuclei when placed in a magnetic field, can absorb energy in the form of radio frequency electromagnetic radiation. You are familiar with the reasons why we have chosen to direct our attention almost entirely to hydrogen. (If you cannot remember then re-read frame 9.) Remember that samples are normally measured in the liquid state or in solution because a sharp absorption band is obtained under these circumstances, and that absorption occurs when the resonance condition ($H_0 = h\nu/2\mu$) is attained.

From what you have learnt so far it would appear that all protons in an organic molecule, regardless of their nature, would give rise to one single absorption. (Think about it!) Reasons why this is not the case are explained in the next section.

PART 2
CHEMICAL SHIFT AND INTEGRATION

Aim

The intention in this section is to give the student a fundamental understanding of two important aspects of nuclear magnetic resonance spectroscopy: chemical shift and integration. In addition to the use of fundamental principles he is encouraged to use n.m.r. data empirically in the interpretation of simple spectra. The units of n.m.r. spectroscopy are covered in detail. On completing the section he should be able to derive useful structural information from spectra, and in simple cases use the technique to make a complete identification of an unknown substance.

New Terms and Concepts

Induced magnetic fields within molecules, diamagnetic shielding and deshielding, chemically and magnetically different protons, integration, parts per million (p.p.m.), δ-values, τ-values, ring current, diamagnetic anisotropy.

Prior Knowledge

Part 1, the structure of simple aliphatic and aromatic compounds, simple electronic effects in organic molecules.

Objectives for Part 2
When you have completed this part you should be able to:
1. Use fundamental principles to recognize situations in molecules where bonding electrons (a) make a negligible contribution to, (b) increase, or (c) decrease the magnetic field at a particular proton.
2. Interpret situations referred to in (1) in terms of shielding or deshielding effects and predict relative positions of absorption along a field scale.
3. Distinguish between chemically different (and therefore magnetically different) protons and chemically identical protons.
4. State the reason why resonance positions of protons are expressed in terms of 'chemical shift' rather than in absolute field terms, and explain in less than 200 words the meaning of this term and why tetramethylsilane is a popular reference substance.
5. Interrelate chemical shifts expressed in hertz to δ-values and τ-values.
6. Use fundamental principles to arrange selected sets of protons in a molecule, which differ in chemical shift by more than 1.0 p.p.m., in the correct order of increasing δ-value.
7. Recognize an 'integration curve' and be able to use it to predict the correct ratio of number of magnetically different protons in a molecule.
8. Recognize protons in a molecule whose chemical shift will depend markedly on conditions in the sample tube.
9. Use a chemical shift correlation table and the principles given in Part 2 to identify the correct structure of an unknown compound, given its n.m.r. spectrum and a set of alternative structures (simple cases only, i.e. no coupling).
10. Deduce a possible structure for a simple unknown compound given its n.m.r. spectrum and a chemical shift correlation table.

Introduction

According to the basic principles outlined in the first part of this programme, the nuclear magnetic resonance spectrum of ethanol C_2H_5OH should consist of a single peak (Fig. 2.1). In fact, it is much more complicated (Figs. 2.2 and 2.3) and, as a consequence, much more informative. There are two features which cause this:

(i) CHEMICAL SHIFT — chemically different protons absorb at different fields.
(ii) PROTON SPIN–SPIN COUPLING — a mutual interaction between neighbouring protons.

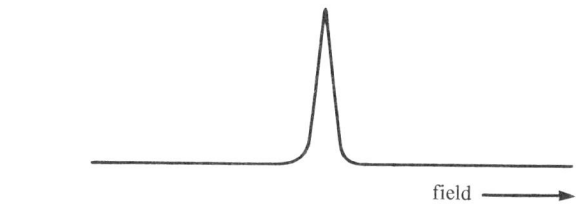

Fig. 2.1.

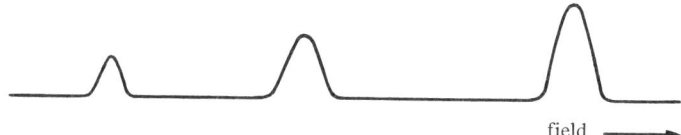

Fig. 2.2. The effect of chemical shift. Ethanol, neat, low resolution.

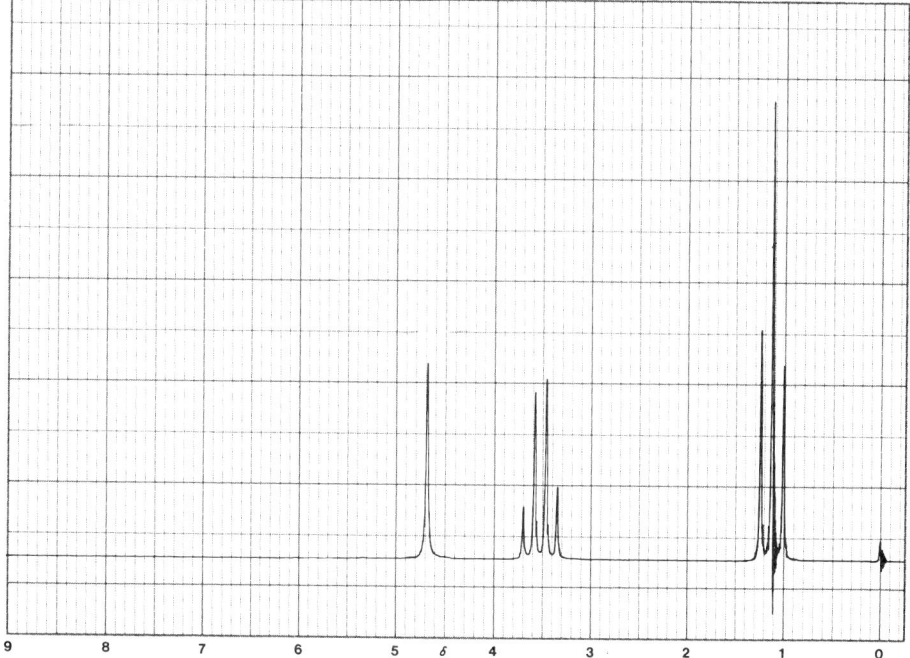

Fig. 2.3. The effect of chemical shift and spin–spin coupling. Ethanol, neat, acidified, high resolution.

A further feature, INTEGRATION, allows the relative numbers of each kind of proton to be ascertained.

In the following section we shall be concerned with CHEMICAL SHIFT and INTEGRATION only, and after completing it, you will be able to derive structural information from spectra such as the one shown in Fig. 2.2. Spin—spin interactions which lead to the ultimate spectrum (shown in Fig. 2.3) will be discussed in Part 3. Although we have tried, in the present section, to avoid examples where spin—spin interaction occurs, some compounds with this feature have been included. In such cases we have deliberately eliminated the fine structure due to the interaction and referred to the results as 'low resolution' spectra. Thus the spectrum in Fig. 2.2 is a low resolution spectrum of ethanol.

As with Part 1, we would like you to undertake the following multiple choice test at this point. Write your answers down on a separate sheet of paper and mark the result in the manner described in Part 1. Do not seek correct answers to incorrectly answered questions unless your mark exceeds seven.

Multiple Choice Test

1. Electrons in σ-bonds e.g. in C—H, exert:
   (a) no effect;
   (b) a shielding effect;
   (c) a deshielding effect
   on the proton which they bind to the rest of the molecule.

2. The molecule **1** has (a) 3, (b) 4, (c) 6 sets of magnetically different protons.

3. $\delta =$ (a) $\dfrac{\text{shift in hertz} \times 10^6}{\text{frequency of instrument in hertz}}$

   (b) $\dfrac{\text{shift in hertz}}{\text{frequency of instrument in hertz} \times 10^6}$

   (c) $10 - \dfrac{\text{shift in hertz} \times 10^6}{\text{frequency of instrument in hertz}}$

4. $CH_3^a - CH_2^b - O - CH_2^c - O - CH_2 - CH_3$
   The δ-values of the protons *a*, *b* and *c* in this compound decrease in the order:
   (a) *a>b>c*; (b) *c>b>a*; (c) *c>a>b*.

5. Select the correct order of decreasing applied field required to bring the following protons into resonance.

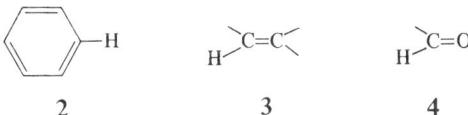

(a) 2>3>4; (b) 3>2>4; (c) 4>3>2.

6. Which of the following statements is **untrue**? Tetramethylsilane is a good reference in n.m.r. spectroscopy because:
   (a) it gives a strong sharp absorption signal due to its twelve equivalent protons;
   (b) its chemical shift is well separated from other absorptions encountered in organic molecules, appearing at exceptionally low field;
   (c) it is chemically unreactive.

7. The n.m.r. spectrum of compound **5** would be expected to show:

$$\begin{array}{c} CH_3 \\ \end{array} C=C \begin{array}{c} H \\ CH_2CH_2CH_2 \end{array} C=C \begin{array}{c} OCH_3 \\ CH_3 \end{array}$$

5

(a) 5 absorption bands of relative intensity 3:3:1:2:2;
(b) 4 absorption bands of relative intensity 6:1:2:1;
(c) 5 absorption bands of relative intensity 3:3:1:2:1.

8. A γ-lactone ($C_8H_{11}O_2$) gives the n.m.r. spectrum shown in Fig. 2.4. Use Chart 1 to deduce whether its structure is (a) **6**, (b) **7** or (c) **8**.

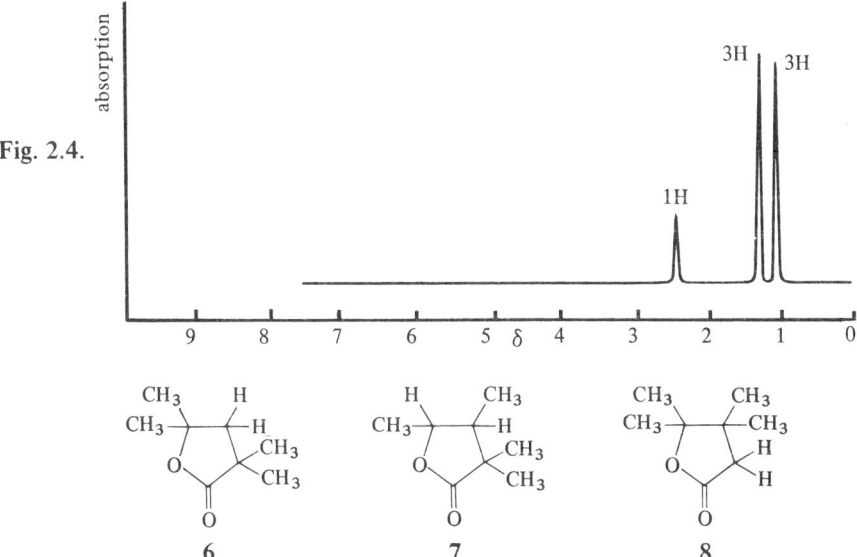

Fig. 2.4.

9. Which of the three statements below is **true**?

The chemical shift of NH protons varies according to conditions of measurement because (a) nitrogen has an odd atomic number; (b) of the influence of hydrogen bonding effects; (c) of the special magnetic properties of the nitrogen nucleus.

10. The low resolution n.m.r. spectrum of one of the isomers of trichloropropane, known to be either (a) $CH_3-CHCl-CHCl_2$, (b) $CH_3-CCl_2-CH_2Cl$, or (c) $CH_3-CH_2-CCl_3$ is shown in Fig. 2.5. Use Chart 1 to identify the compound.

Fig. 2.5.

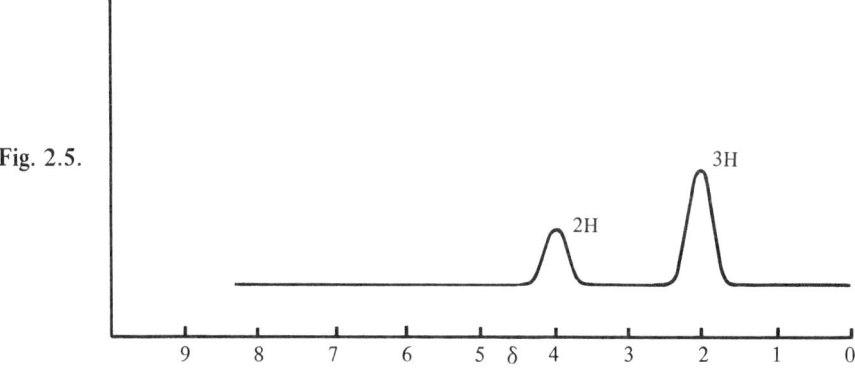

Comments on marks

Less than 2: Continue with Part 2.

3–7: You have some knowledge of this topic but it would be in your interest to secure a more thorough understanding and you are advised to work through Part 2 – it shouldn't take you long!

More than 7: Unless you obtained full marks, consult the sections of Part 2 relevant to your incorrect answers; then proceed directly to Part 3.

PART 2: CHEMICAL SHIFT AND INTEGRATION

Diamagnetic Shielding

F30 The field H in which radiation can induce proton magnetic resonance $(H = h\nu/2\mu)$ is that which is present at the precise location of that particular proton. Up until now we have assumed this to be the *same as/different from* the applied field.

F31 In fact the field at the site of a proton in a molecule is influenced by nearby electrons. In the light of this new knowledge were we justified in our assumption that the strength of the field influencing the proton was identical with that of the applied field?

A30 same as.

F32 The proton is bound to the rest of the molecule by a two-electron bond, the electrons of which are induced by an applied magnetic field to circulate in a plane perpendicular to the field. Fill in the missing words to complete Fig. 2.6 which can be used to illustrate this simplified but nevertheless useful picture.

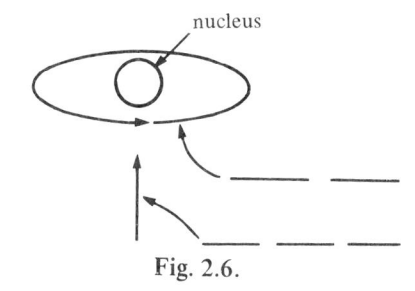

Fig. 2.6.

A31 No. The field at the nucleus is different from the applied field. The following frames indicate how this difference arises.

F33 Now electron circulations give rise to magnetic fields. Thus the effect of an applied magnetic field on the electrons of a molecule is to produce localized **induced** magnetic fields. Moreover, these induced magnetic fields operate against the applied field at the proton. Complete Fig. 2.7.

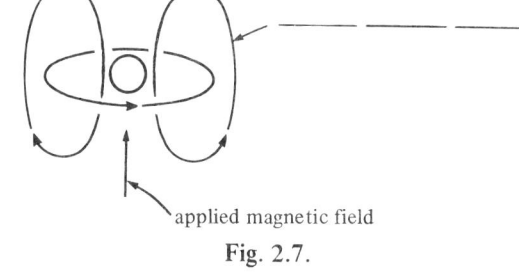

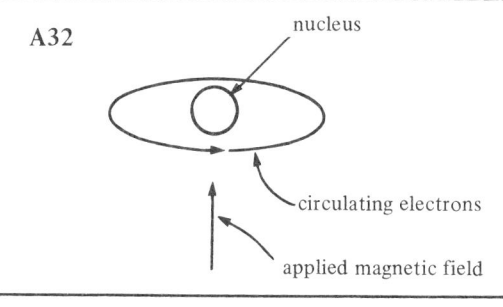

Fig. 2.7.

A32

Fig. labels: nucleus, circulating electrons, applied magnetic field

F34 Consider your diagram in answer 33 and answer the following question. Which would you expect to require the highest applied field for resonance? (a) A proton with valence electrons. (b) An 'isolated' proton.

A33 *(diagram: induced magnetic field around proton, with applied magnetic field)*

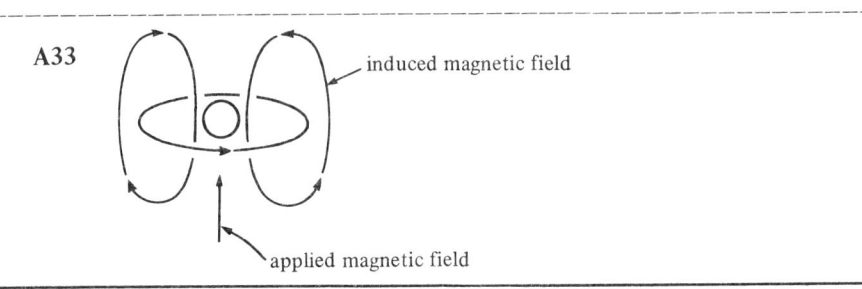

F35 Let the field at which an isolated proton (i.e. without valence electrons) resonates be H_0 and the induced field created by bonding electrons experienced by a proton attached to a molecule be H' (see Fig. 2.8). What field, H_{app}, would we have to apply externally to achieve the resonance condition of the proton bound to the molecule (in terms of H_0 and H')?

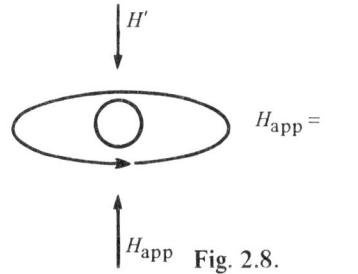

Fig. 2.8.

A34 (a) A proton with valence electrons. A greater applied field is needed to overcome the 'shielding' effect of the induced magnetic field of the valence electrons.

F36 Thus, in order to attain the resonance condition, a proton with associated electrons requires a greater applied field ($H_0 + H'$), than that required by an isolated proton (H_0). We say that the electrons **shield** the proton from the applied magnetic field, and the phenomenon is frequently referred to as DIAMAGNETIC SHIELDING. A proton under the influence of diamagnetic shielding will resonate at *higher/lower* applied field than an unshielded proton.

A35 $H_{app} = H_0 + H'$

F37 The spectrum in Fig. 2.9 is that of a compound with two sets of protons. Which is the more highly shielded, A or B?

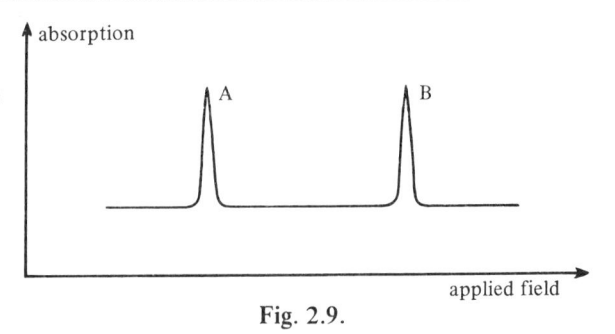

Fig. 2.9.

A36 higher.

F38 As shielding arises from a magnetic field created by induced electron circulation, it is not surprising that its effect varies with the electron density in the neighbourhood of the proton.

Which would you expect to be more highly shielded, a nucleus in an area of low electron density, X: H, or one in an area of high electron density, X :H?

A37 B. B absorbs at higher field strengths than A.

F39 You should now be able to see why all protons in organic molecules do not resonate at the same applied field. In fact as valence electrons which determine the chemical properties of the X–H bond are so important in shielding, we can say that chemically different protons are also magnetically different; that is they will resonate at different fields in a nuclear magnetic resonance spectrometer. How many different absorptions would you see in the spectrum of propan-1-ol ($CH_3CH_2CH_2OH$)?

A38 Greatest shielding is expected in an area of high electron density X :H.

F40 The bond polarizations discussed in the previous two frames and caused by differing electronegativity of neighbouring atoms is known as the INDUCTIVE EFFECT.

X: H or X←H

It is relayed through the sigma (σ) bonds of a molecule, and its magnitude falls off rapidly with distance from the electronegative atom. Thus, in ethanol we have the situation shown in compound **9**.

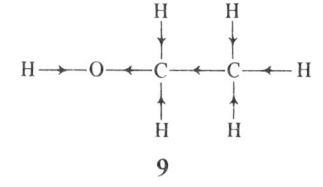

H→—O←—C←—C←—H (with H substituents)

9

Is the inductive effect of oxygen felt more strongly at the methylene protons, or methyl protons of ethanol?

A39 Four. There are four chemically different types of hydrogen in the molecule — methyl, two different methylenes and hydroxyl.

F41 Which of the three types of proton in ethanol will be most effectively shielded by bonding electrons?

A40 At the methylene protons.

F42 The three types of proton in ethanol absorb at different field strengths. On the basis of electronegativity differences, predict the order in which they resonate in an increasing field.

A41 Methyl. The methyl protons are furthest from the electronegative oxygen atom.

F43 Figure 2.10 is a low resolution* n.m.r. spectrum of ethanol.

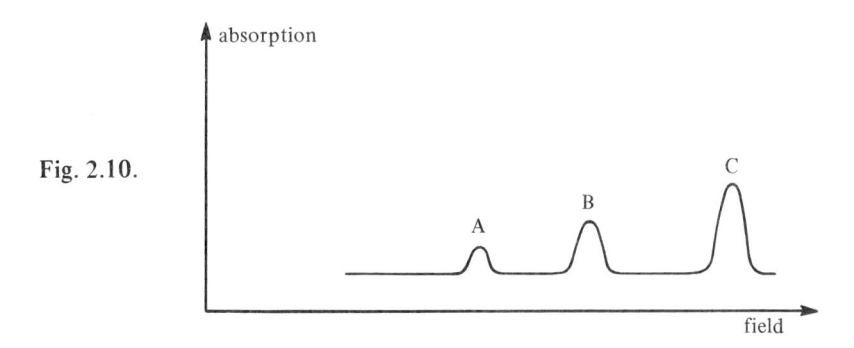

Fig. 2.10.

Which of the absorptions, A, B or C arise from: (a) the methylene protons, (b) the methyl protons, (c) the hydroxyl proton?

A42 $-OH$, $-CH_2-$, $-CH_3$.

F44 Figure 2.11 shows a low resolution spectrum of 1-chloro-2-iodoethane, $ClCH_2CH_2I$.

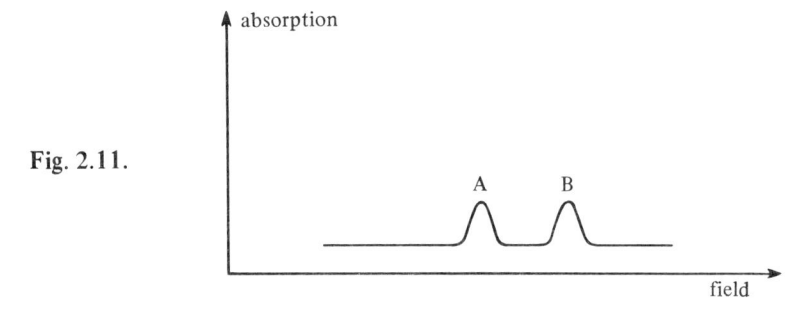

Fig. 2.11.

Which of these signals corresponds to the hydrogens of the methylene group bonded to iodine?

A43 (a) B, (b) C, (c) A.

F45 You recall that **chemically** different protons are also **magnetically** different, and, as a consequence give rise to nuclear magnetic resonance absorptions at different fields.

*As mentioned in the introduction to Part 2, a low resolution spectrum is one in which the fine structure of absorption bands is not evident.

The n.m.r. spectra of dimethyl ether, CH₃OCH₃ and methyl acetate CH₃COOCH₃ are given in Fig. 2.12. Identify the methyl acetate spectrum.

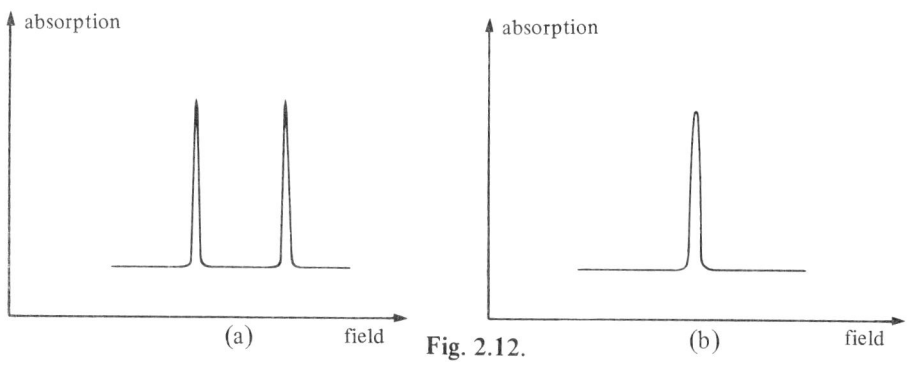

Fig. 2.12.

A44 B. Iodine is less electronegative than chlorine. Therefore the –CH₂I protons will be more effectively shielded by C–H bond electrons than will the –CH₂Cl protons.

Note: Although electronegativity exerts a strong influence on field position, other factors also play a part. These become dominant in cases where electronegativity differences are small, for example between methine, methylene and methyl protons. Such factors lead to small but significantly different field positions but the theory of such differences is beyond the scope of the present discussion.

F46 Figure 2.13 is a low resolution n.m.r. spectrum of one of the three compounds below. Which one?

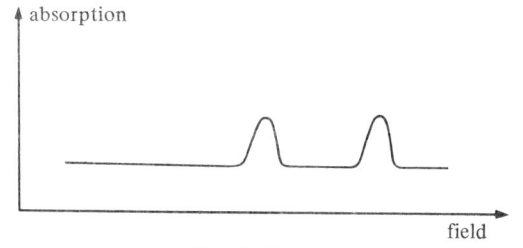

Fig. 2.13.

(a) (CH₃)₃CBr   (b) BrCH₂OCH₂CH₂Cl   (c) $\begin{matrix} H_2C-CH_2 \\ H_2C \quad CH_2 \\ \quad O \end{matrix}$

A45 (a). The protons on the methyl groups of dimethyl ether are chemically identical. One methyl group of methyl acetate is attached to carbon and the other to oxygen. Chemically, then, they are different.

A46 (c). The four protons of the methylene groups adjacent to the oxygen atom are identical, but different from the four identical protons on the other two carbon atoms. All nine of the protons in (a) are identical and would give rise to one absorption; (b) has three pairs of chemically different protons.

Integration

F47 Take a look at the spectrum of ethanol shown in Fig. 2.10. You will observe that the sizes (or intensities) of the three signals are different. Given that the area under an n.m.r. signal is directly proportional to the number of protons responsible for it, what is the area ratio of the three absorptions in this spectrum?

F48 In practice the area under an n.m.r. signal is measured electronically by the spectrometer and superimposed on the spectrum in the form of a step in a horizontal line, the height of which is proportional to the area under the absorption. This measurement is known as INTEGRATION. Thus a low resolution spectrum of ethanol, complete with its integration curve would look like Fig. 2.14.

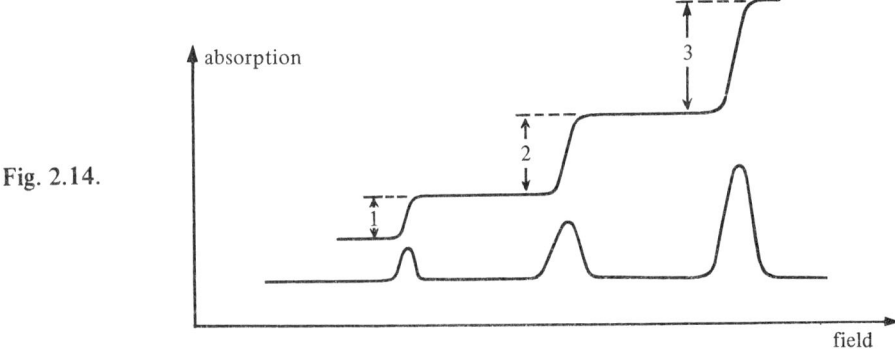

Fig. 2.14.

Examine the integration curve in the spectrum in Fig. 2.15 and, with the use of a ruler, preferably with a millimetre scale, determine the area ratio of the two signals.

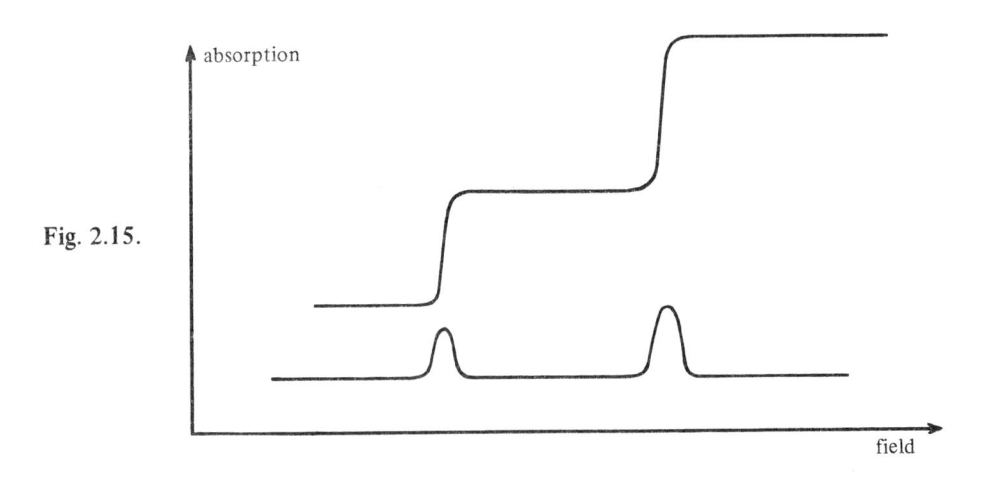

Fig. 2.15.

A47 A:B:C; 1:2:3.

F49 Integration is particularly valuable in the analysis of n.m.r. spectra. In fact, in the case of a compound C_2H_4O, we can derive its structure from this feature alone.

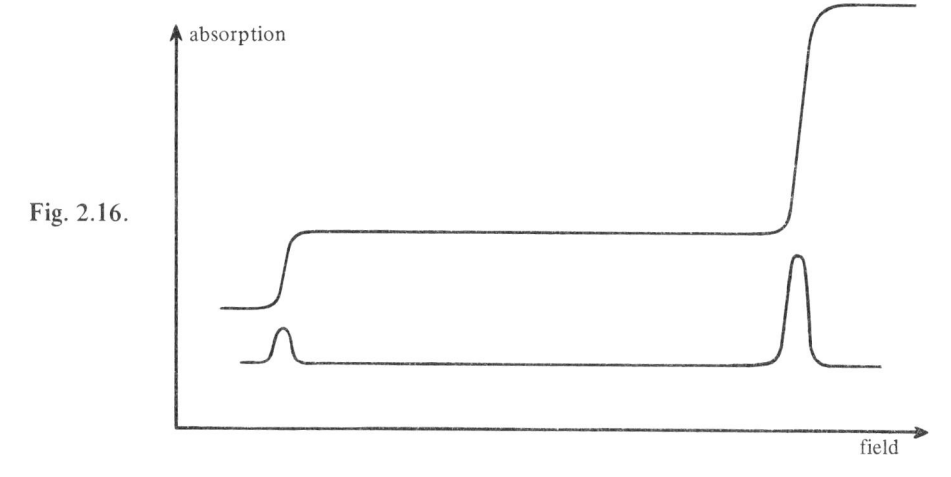

Fig. 2.16.

Its low resolution spectrum is shown in Fig. 2.16 together with an integration curve. What is its structure?

A48 Area ratio 3:4.

F50 Whilst in most cases we need much more information than integration to effect a complete analysis, it is often sufficient to distinguish between a small number of alternatives. For example the spectrum in Fig. 2.17 is that of a compound known to be either:

(a) $CH_3COCH_2COOC(CH_3)_3$, (b) $CH_3CH_2COCH_2CH_2COCH_2CH_3$,

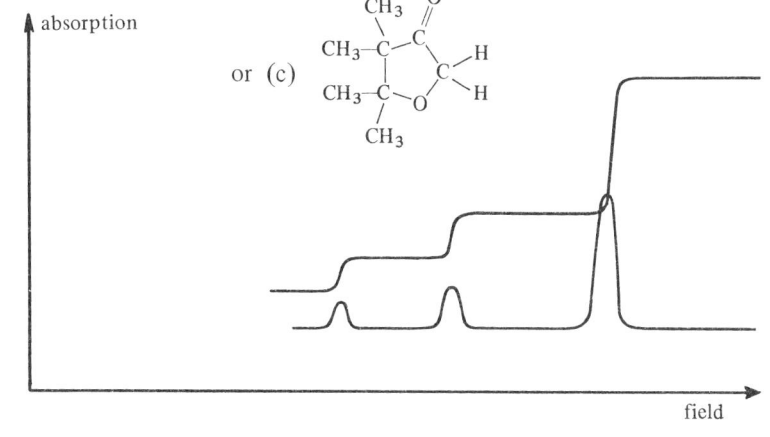

or (c)

Fig. 2.17.

Which one is consistent with the spectrum?

A49 CH₃CHO. The area ratio is 1:3. The only structure consistent with this and the molecular formula is CH₃CHO.

F51 Consider the compound 2-chloro-3-iodopentane (**10**).

$$\overset{1}{CH_3}-\overset{2}{CH}-\overset{3}{CH}-\overset{4}{CH_2}-\overset{5}{CH_3}$$
$$||$$
$$ClI$$

10

How many signals would you expect to see in a low resolution n.m.r. spectrum?

A50 (a) CH₃COCH₂COOC(CH₃)₃. (b) would give a spectrum with absorptions of area ratio 2:2:3 and (c) 1:3:3. The area ratios in the spectrum are 2:3:9.

F52 Which of the two methyl groups of 2-chloro-3-iodopentane would you expect to occur at highest field?

A51 Five.

F53 Which of the two methine hydrogens of 2-chloro-3-iodopentane would you expect to occur at lowest field?

A52 $-\overset{5}{CH_3}$. These protons are furthest away from the electronegative halogen atoms and as such would be expected to be the most shielded.

F54 What will be the ratio of heights of the steps on the integration curve on the spectrum of 2-chloro-3-iodopentane for the protons on:
$\overset{1}{C}, \overset{2}{C}, \overset{3}{C}, \overset{4}{C}$ and $\overset{5}{C}$?

A53 $-\overset{2}{\underset{|}{\underset{Cl}{CH}}}-$ The proton is closest to the most electronegative atom in the molecule.

F55 Draw the low resolution n.m.r. spectrum which you would expect to obtain from 2-chloro-3-iodopentane given that both methyl groups absorb at higher fields than the methylene group.

A54 3:1:1:2:3.

F56 Figure 2.18 shows the low resolution n.m.r. spectrum of a compound of molecular formula, C_3H_8O. Let us see what steps are involved in establishing its structural formula.

First what is the ratio of protons in the three absorptions?

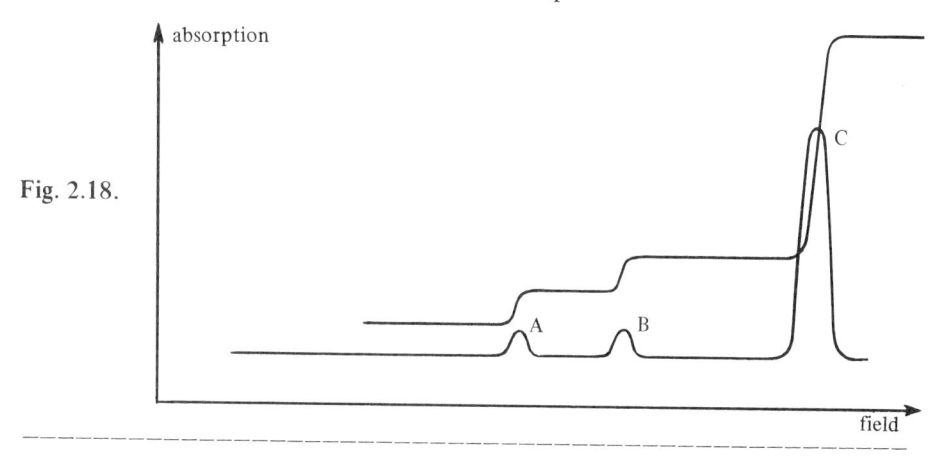

Fig. 2.18.

A55 This spectrum (Fig. 2.19) takes into account the relative shifts expected on the field scale and the expected area ratios. You are only expected to get the order and relative sizes of the peaks correct.

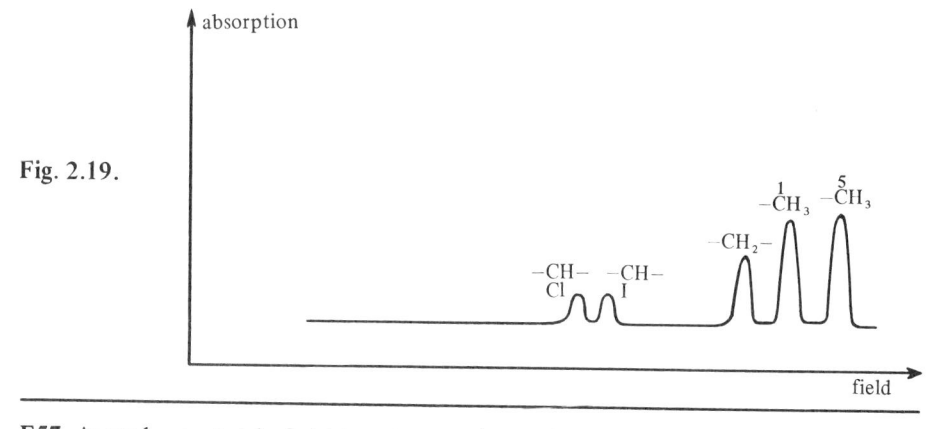

Fig. 2.19.

F57 As we have a total of eight protons we know that the absorptions A, B and C correspond to 1, 1 and 6 protons respectively. (For 2, 2 and 12 we would need 16 protons in the molecule.)

Consider absorption C. This requires (a) six identical methine protons or (b) three identical sets of methylene protons or (c) two identical methyl groups. Look at the molecular formula and you will see that only one of these possibilities can be entertained. Which one?

A56 1:1:6.

F58 Two methyl groups incorporate six protons and two carbon atoms; we therefore have only one carbon left. Oxygen can be present as a part of an ether, carbonyl or hydroxyl group. With this in mind, only one structure fits all the data. Write it down.

A57 (c) two identical methyl groups. (a) is impossible because it requires at least six carbon atoms and (b) is impossible as one cannot write a structure which satisfies all the data.

F59 Deduce the structural formula of the compound C_3H_8O whose low resolution n.m.r. spectrum is given in Fig. 2.20, and assign each signal to a group of protons.

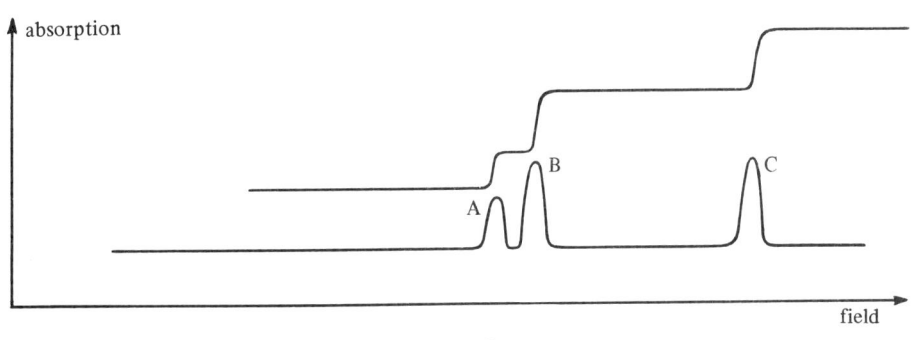

Fig. 2.20.

A58

$$\begin{array}{c} H \\ | \\ CH_3-C-CH_3 \\ | \\ OH \end{array}$$

11

A59 $\overset{B}{CH_3}-O-\overset{A}{CH_2}-\overset{C}{CH_3}$. The first thing to note is that there are three different types of protons. The absorptions A, B and C have area ratios 2:3:3 respectively and the molecular formula indicates that they correspond to $-CH_2$, $-CH_3$ and $-CH_3$. With only one oxygen atom remaining the compound must be ethyl methyl ether.

Chemical Shift and its Units

F60 Up to this point, we have been able to talk about the fields at which resonance occurs for various protons in a purely qualitative and relative sense, for example 'A occurs at higher fields than B'. If we were able to record some field value at which a particular type of proton resonates, it would be very helpful in recognizing that type of proton. Although the field is being swept, it is not practical for us to use **absolute** field values for the following reason. The fields employed in n.m.r. spectrometers are enormous (e.g. 14 000 G for a 60 MHz instrument). By comparison, the range swept is minute. The situation is comparable to an attempt to measure the weight of a man on an ocean liner by the difference in weight of the vessel when he is on board and when he is not. Accordingly, field positions of absorptions are given not in absolute units, but relative to the absorption position of some standard.

So far we have been able to derive information from a spectrum using three factors:

1. Chemically different protons are magnetically different and give rise to absorptions at different field strengths.
2. The integration curve gives the ratio of magnetically different types of protons.
3. The greater the shielding of a proton by its valence electrons, the higher the field at which it resonates.

From the information in this frame, a fourth factor becomes evident. What is it?

F61 By far the most common standard used is tetramethylsilane, $(CH_3)_4Si$ (frequently abbreviated to TMS) in which all protons are equivalent. Recalling that silicon is less electronegative than carbon, would you expect the protons of $(CH_3)_4Si$ to be more effectively shielded and hence at higher field than those of ethane?

    A60 A numerical value may be given to the position of the absorption signal.

F62 In fact the protons of TMS are more highly shielded than virtually all other protons which we are likely to meet in organic compounds. Would you expect the TMS signal to be on the right hand side, in the middle, or on the left hand side of the spectrum of an organic substance to which it has been added, assuming the field increases from left to right?

    A61 Yes. The electron density at the protons in TMS will be greater than that in the protons in ethane. TMS protons will thus be more effectively shielded and resonate at higher field.

F63 The signals of most protons in organic compounds are observed at lower field than TMS. The distance from the TMS signal of a signal from a sample of an organic compound is known as its CHEMICAL SHIFT. N.m.r. signals from most protons in organic molecules have chemical shifts *up-field/down-field* from TMS.

    A62 The TMS signal is found on the right hand side of the chart which corresponds to the largest magnetic fields.

F64 You recall that the horizontal scale in an n.m.r. spectrum is one of field. It has been found more satisfactory, however, to express chemical shifts in frequency units, hertz (Hz), instead of a difference in magnetic field values (gauss). This is perfectly straightforward as we know that field and frequency are directly proportional to each other:

$$\nu = \frac{2\mu H}{h} \quad \text{(i.e. } \nu = \text{constant} \times H\text{)}$$

Fig. 2.21.

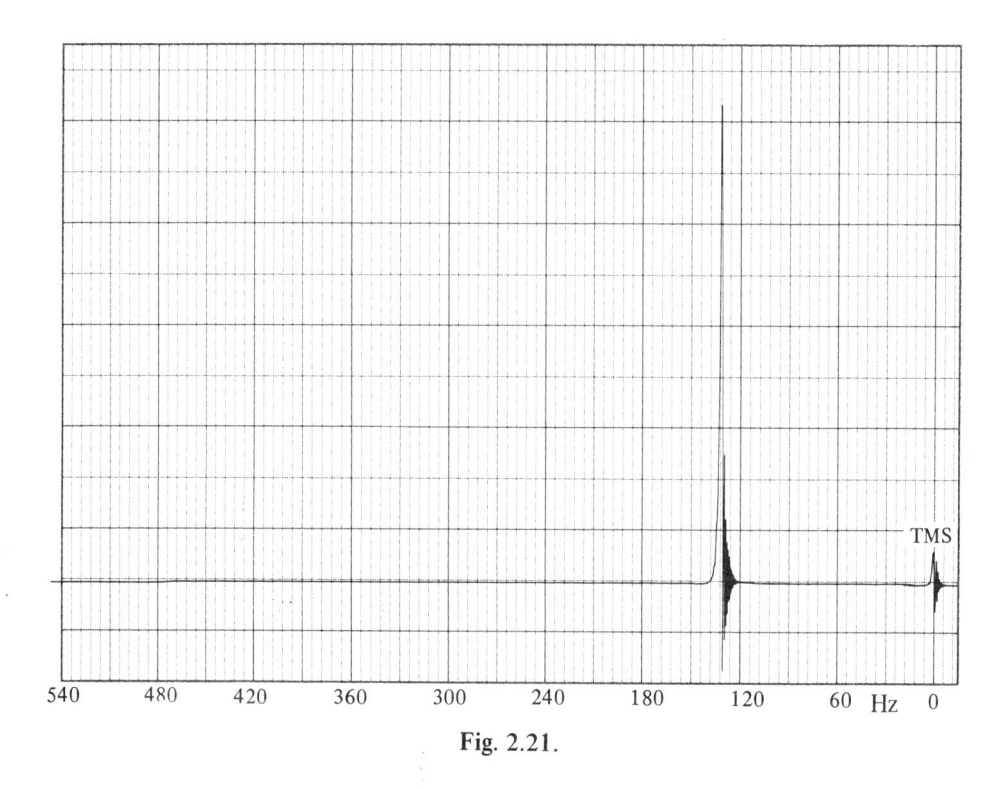

In fact one never has to make the calculation as the chart paper on which n.m.r. spectra are recorded is already calibrated in hertz. A typical n.m.r. spectrum of acetone CH_3COCH_3 measured on an instrument operating at 60 MHz is shown in Fig. 2.21. What is the chemical shift of the methyl protons of acetone in hertz?

    A63 down-field.

F65 In practice one either uses a solvent containing a small amount of TMS, or adds a drop of TMS to the solution of the sample before insertion between the poles of the spectrometer magnet. TMS is chemically inert, and only interferes with the n.m.r. spectrum of a sample in the exceptionally rare instance of a particular coincidence. What would be the nature of the interference in this case?

    A64 130 Hz. Note that from now on all spectra will be presented with a
    horizontal scale in frequency units.

F66 We have already indicated that not all instruments operate at the same field (and consequently, the same frequency). It turns out that chemical shift is directly proportional to the applied field (and frequency). Suppose we measure the n.m.r. spectrum of acetone on two instruments, A and B, and the magnetic field of B is twice that of A. Is the chemical shift of acetone measured with B, half or twice that with A?

A65 When a signal from the sample overlaps with or is hidden by the TMS signal.

F67 The spectrum of acetone in frame 64 is measured with an instrument operating at 60 MHz. What would be the chemical shift when measured with a 100 MHz instrument?

    A66 Twice.

F68 Clearly it is desirable that chemical shifts be measured in units which are independent of spectrometer frequency so that we can say, for example, 'The chemical shift of acetone from TMS is X' without having to specify the instrument on which the measurement was made. This is achieved by dividing the chemical shift in hertz by the spectrometer frequency in hertz. Most commercial instruments at the present time operate at 60 or 100 MHz. The spectrum of acetone shown in frame 64 was measured with a 60 MHz instrument. What would be the chemical shift of the methyl protons in the new unit which is independent of spectrometer frequency?

    A67 216.7 Hz. Chemical shift when measured with a 100 MHz instrument

$$= 130 \times \frac{100}{60} \text{ Hz.}$$

F69 You notice that the number given in answer 68 is very small. For convenience this is multiplied by 10^6 to make the numbers easier to handle. The unit of chemical shift is now referred to as 'parts per million' and is given the symbol δ (delta).

i.e. $\delta = \dfrac{\Box}{\Box} \times \Box$

Insert (a), (b) and (c) into the appropriate boxes:

(a) shift of signal from TMS in hertz;
(b) spectrometer frequency in hertz;
(c) 10^6.

    A68 2.17×10^{-6}

$$\frac{130 \text{ Hz}}{60 \text{ MHz}} = \frac{130}{60 \times 10^6}$$

F70 What is the chemical shift of acetone in p.p.m. from TMS in the spectrum shown in Fig. 2.21?

    A69

$$\delta = \frac{\boxed{a}}{\boxed{b}} \times \boxed{c}$$

F71 The further down-field from TMS a signal occurs, the *higher/lower* the δ-value.

A70 2.17 p.p.m. $= \dfrac{130}{60 \times 10^6} \times 10^6$ p.p.m.

F72 One criticism of δ is that the **greater** the field at which a signal occurs, the **smaller** the unit of chemical shift — a potential source of misunderstanding. Accordingly some chemists use the τ (tau)-unit which is related to δ by the simple expression:

$$\tau = 10 - \delta$$

and TMS is given the value 10.

Why would a signal with a τ-value of 12.6 be unusual in organic compounds?

A71 higher. The further down-field from TMS the greater the chemical shift in hertz. P.p.m. is directly proportional to hertz.

F73 What is the τ-value of the protons in acetone according to the spectrum in Fig. 2.21.

A72 The protons giving rise to the signal would be resonating at higher field than those of TMS (*cf.* frame 63).

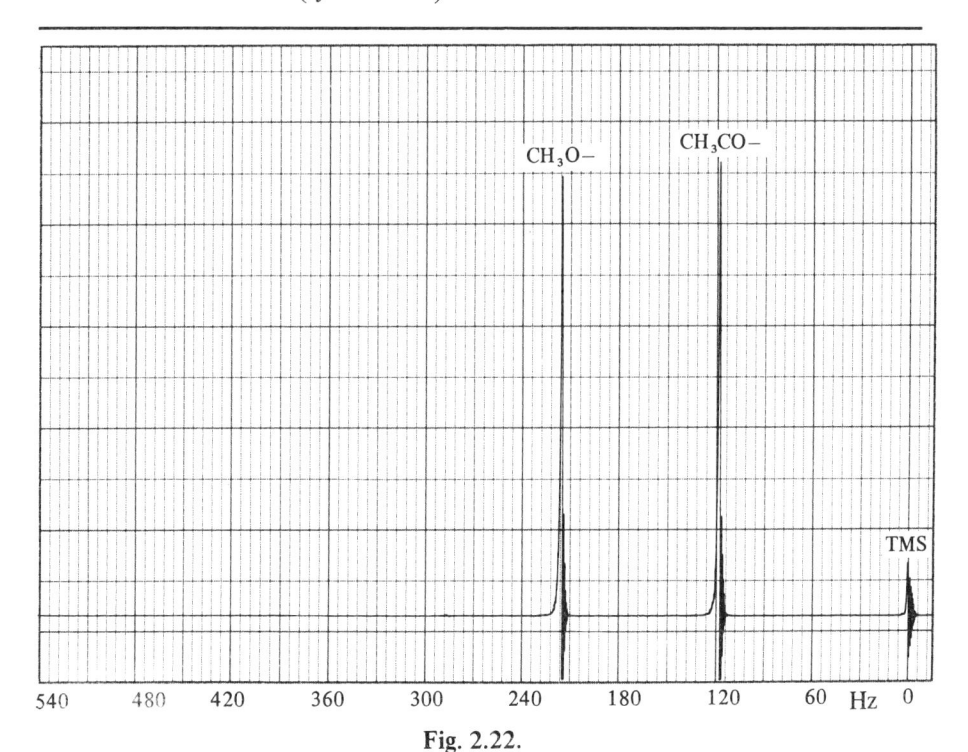

Fig. 2.22.

F74 Figure 2.22 shows the spectrum of methyl acetate measured on an instrument operating at 60 Hz. Determine the δ-values for the two different types of protons.

A73 7.83 τ (10 – 2.17). The τ-scale also has disadvantages (e.g. the greater the chemical shift the smaller the value) and is currently less favoured than the δ-scale. Accordingly we shall use the latter throughout the rest of the programme.

F75 We mentioned earlier that there is a reasonable correlation between shielding characteristics and electronegativities of substituents in organic molecules. You should now be able to express this quantitatively by selecting the correct set of δ-values for the following series. Values for the protons of the methyl halides, CH_3I, CH_3Br, CH_3Cl, CH_3F are:

(a)	4.26	3.05	2.68	2.16	p.p.m. respectively
(b)	4.26	2.68	2.16	3.05	p.p.m. respectively
(c)	2.16	2.68	3.05	4.26	p.p.m. respectively

A74 CH_3O-, 3.6 p.p.m.; CH_3CO-, 2.0 p.p.m. (δ-Scales are used from now on.)

F76 Similarly the cumulative effect of increasing the number of substituent halogens may be seen by assigning the correct δ-values to the proton resonances of CH_3Cl, CH_2Cl_2 and $CHCl_3$. Which sequence is correct?

(a)	5.30	3.05	7.27	p.p.m. respectively.
(b)	3.05	5.30	7.27	p.p.m. respectively.
(c)	7.27	5.30	3.05	p.p.m. respectively.

A75 (c). The electronegativities of the halogen atoms increase in the order given, consequently the shielding of the methyl protons decreases in the order given. It follows that the δ-values increase with increasing electronegativity of the halogen atom.

F77 Methylene protons of the type $R^1OCH_2R^2$, where R^1 and R^2 are alkyl groups, have a δ-value in the range 3.4 to 4.4 p.p.m. Which of the values 5.3 or 3.0, would you expect for methylene protons in a molecule of the type 12?

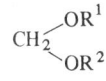

$$CH_2\begin{matrix}OR^1\\OR^2\end{matrix}$$

12

A76 (b). The electron density at the protons of these three compounds decreases as hydrogen is progressively replaced by the more electronegative chlorine.

F78 A carboxylic acid proton RCOOH has a chemical shift of 11. Would this be a δ- or a τ-value?

A77 5.3 p.p.m. Two electronegative oxygen atoms reduce the electron density at the methylene proton to a greater extent than one.

F79 From the exercises in the previous few frames, it is clear that with a little knowledge of the theory of n.m.r., we can predict trends of δ-values between similar compounds. In practice, we generally recognize particular types of protons in an n.m.r. spectrum by the use of empirical knowledge. A vast amount of chemical shift data has been collected in recent years from which it is clear that the δ-values of protons in particular environments in organic molecules do not vary appreciably from molecule to molecule. For example, the signal at about 2.2 p.p.m. for the methyl protons of acetone is a characteristic of the spectra of all methyl ketones. Thus we can classify protons **according** to their δ-value and, conversely, we can recognize different types of protons **by** their δ-values. There are many compilations of chemical shift values available in the literature (a bibliography can be found at the end of this section). A useful list is given in the pull-out sheet (Chart 1) to be found at the end of the programme. Pull it out and keep it out. The chart is made up of four sections. The first three give chemical shifts for methyl, methylene and methine protons, and the fourth for protons attached to multiple bonds.

The table may be used in an empirical way to predict the chemical shifts of protons in a variety of environments. For example, suppose one wished to predict the chemical shifts of the protons of the groups a and b in 2,2-dimethylbutane (**13**).

$$CH_3-CH_2-\underset{\underset{CH_3}{|}}{\overset{\overset{CH_3}{|}}{C}}-CH_3$$
$$ab$$

13

According to the chart, the methyl group a is classified as being adjacent to $-CR_3$ where $R_3 = H, H$ and $\overset{|}{C}-$. Line 1 gives the chemical shift range for such groups as 0.85 to 1.1 p.p.m. The methylene group b, $R-CH_2-$, is also adjacent to a $-CR_3$ group, where this time $R_3 = (CH_3)_3$. Line 9 of the chart shows that these protons will absorb within the range 1.2 to 1.7 p.p.m.

In the related chloro compound 2-chloro-2-methylbutane (**14**) the methyl group c is in a similar environment to a and will absorb between 0.85 and 1.1 p.p.m. Use Chart 1 to predict the chemical shift range for the methylene group protons d.

$$CH_3-CH_2-\underset{\underset{CH_3}{|}}{\overset{\overset{CH_3}{|}}{C}}-Cl$$
$$cd$$

14

A78 δ. The electron density at the proton of a carboxyl group is relatively low and so the δ-value will therefore be relatively high.

F80 Use Chart 1 to determine the chemical shift ranges for the protons in propyl benzoate (**15**) and complete the table.

$$\text{C}_6\text{H}_5-\overset{\text{O}}{\underset{a}{\text{C}}}-\text{O}-\underset{b}{\text{CH}_2}-\underset{c}{\text{CH}_2}-\underset{d}{\text{CH}_3}$$

15

Protons	Adjacent Group	δ-Range
Aryl (a)	Not applicable	6.6 to ?
Methylene (b) R–CH$_2$–	?	?
Methylene (c) R–CH$_2$–	CR$_2$X where R$_2$ = H, H and X = –O–	?
Methyl (d) CH$_3$–	? where R = ?	0.85 to 1.1

A79 1.2 to 1.95 p.p.m. R–CH$_2$ in 2-chloro-2-methylbutane is adjacent to CR$_2$X where R$_2$ = (CH$_3$)$_2$ and X is Cl (an electron withdrawing atom). The chemical shift range for an R–CH$_2$–CR$_2$X methylene group is given in the chart on line 10.

F81 Use Chart 1 to give ranges of δ-values for the different types of protons in compounds **16** to **18**.

16: H$_3^b$C–C$_6$H$_4$(H^a)$_4$–CH$_3^b$ H^a = H^b =

17: structure with H^a, CH$_3^b$, O–CH$_3^b$, O, O   H^a = H^b =

18: CH$_3^a$–CH$_2^b$–O–H   H^a = H^b =

A80 Protons	Adjacent Group	δ-Range
Aryl (a)	Not applicable	6.6 to 8.5 (line 27)
Methylene (b) R–CH$_2$–	–O–	3.35 to 4.4 (line 16)
Methylene (c) R–CH$_2$–	CR$_2$X where R$_2$ = H, H and X = –O–	1.2 to 1.95 (line 10)
Methyl (d) CH$_3$–	CR$_3$ where R = H, H and C–	0.85 to 1.1 (line 1)

F82 You will notice we have not included in our table precise values for protons bound to nitrogen and oxygen. These values are very sensitive to the degree of intramolecular and intermolecular hydrogen bonding, which, in turn, is critically dependent on concentration, temperature and the nature of the solvent. Consequently hydroxylic protons, for example, may occur almost anywhere in the spectrum. The spectroscopy of protons attached to nitrogen is further complicated by the magnetic and electric properties of the nitrogen nucleus, although these affect the shape of the signal rather than its chemical shift. Which of the following classes of compounds will contain protons whose chemical shift will depend largely on the conditions in the sample tube? Aldehydes, primary amides, tertiary amines, alcohols, esters, aromatic ethers, secondary amines.

A81
16: H^a 6.6 to 8.5 p.p.m. (line 27); H^b 2.2 to 3.0 p.p.m. (line 6)
17: H^a 5.8 to 7.7 p.p.m. (line 26); H^b 3.25 to 4.3 p.p.m. (line 8)
18: H^a 0.85 to 1.9 p.p.m. (line 2); H^b 3.35 to 4.4 p.p.m. (line 16)

F83 What three variables in **conditions** determine the precise δ-value for the hydroxyl proton resonance of ethanol?

A82 Primary amides, alcohols and secondary amines. Only these compounds contain hydrogen attached to oxygen or nitrogen.

F84 Is the following statement true or false?
An additional feature of hydroxyl proton n.m.r. signals is that they are influenced by the magnetic properties of the oxygen nucleus.

A83 Solvent, concentration and temperature. See frame 82.

A84 False. The case given is for N–H protons.

The Empirical Use of Chemical Shift and Integration
F85 With a knowledge of:
(i) the value and use of integration;
(ii) the fact that most chemically identical protons are magnetically equivalent;
(iii) the influence of the inductive effect on chemical shift; and now
(iv) empirical δ-values for different types of protons;

we are in a position to use n.m.r. to solve problems which involve the distinction between simple structural alternatives and in some simple cases to effect a complete identification.

A compound suspected of being either butanal (butyraldehyde), $CH_3CH_2CH_2CHO$, or pentan-2-one, $CH_3CH_2CH_2COCH_3$ has n.m.r. absorption in the following regions:

1.0, 1.5, 2.4 and 2.7 p.p.m. Identify the compound and assign the signals to the appropriate protons.

F86 The n.m.r. spectrum of a compound suspected of being either **19** or **20** had a number of signals in the region 0.5 to 2.0 p.p.m. but none with a δ-value greater than 3.0 p.p.m. Identify the compound.

$$CH_3CH_2CH\begin{subarray}{c}CH_2\\ |\\ CH_2\end{subarray} \qquad CH_3CH_2CH=CHCH_3$$

19   20

A85 Pentan-2-one (**21**).

$$\underset{1.0}{H_3C}-\underset{1.5}{CH_2}-\underset{2.7}{CH_2}-\overset{\overset{O}{\|}}{\underset{2.4 \text{ p.p.m.}}{C}}-CH_3$$

21

Note, butanal would have a signal due to the aldehyde proton near 9.7 p.p.m. This is well removed from the chemical shift values given.

F87 An unknown organic compound ($C_9H_{20}O$), believed to have either structure **22** or **23** gives the n.m.r. spectrum shown in Fig. 2.23. Which of the two structures can be eliminated on the basis of the number of signals?

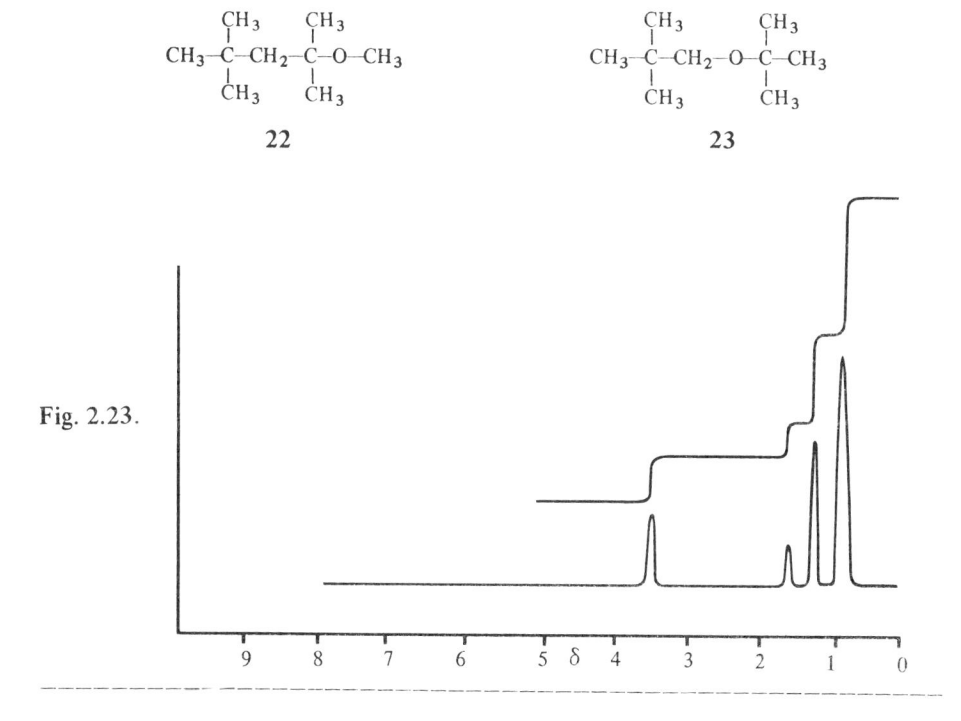

$$CH_3-\underset{\underset{CH_3}{|}}{\overset{\overset{CH_3}{|}}{C}}-CH_2-\underset{\underset{CH_3}{|}}{\overset{\overset{CH_3}{|}}{C}}-O-CH_3 \qquad CH_3-\underset{\underset{CH_3}{|}}{\overset{\overset{CH_3}{|}}{C}}-CH_2-O-\underset{\underset{CH_3}{|}}{\overset{\overset{CH_3}{|}}{C}}-CH_3$$

22   23

Fig. 2.23.

A86 CH₃CH₂CH⟨CH₂/CH₂⟩
24

The chart shows that protons attached to double bonds have values greater than 4.6 p.p.m.

F88 In order to confirm or eliminate structure 22 take a closer look at the spectrum; in particular, measure and write down the area ratio of the signals.

A87 23. This compound has three different types of protons.

F89 The area ratios, then, are consistent with structure 22, but are the δ-values? Go to the table and record the expected range of δ-values for the protons of the *t*-butyl group (*a*), the methylene group (*b*), the methyl group (*c*), and the methoxy group (*d*).

$$\begin{array}{c} CH_3 \\ | \\ CH_3-\underset{|}{\overset{|}{C}}-CH_2-\underset{|}{\overset{|}{C}}-O-CH_3 \\ CH_3 \quad CH_3 \end{array}$$
       a   b   c   d

25

A88 Area ratio, 3:2:6:9.

F90 Do we have four signals, of appropriate area ratios, in the spectrum which would correspond to protons with these chemical shifts?

A89 (*a*) 0.85 to 1.1 p.p.m.; (*b*) 1.2 to 1.95 p.p.m.; (*c*) 0.85 to 1.9 p.p.m.; (*d*) 3.25 to 4.3 p.p.m. Note that according to our chart notation, (*b*) and (*c*) are classified as groups attached to $-CR_2X$ where X in this case is oxygen.

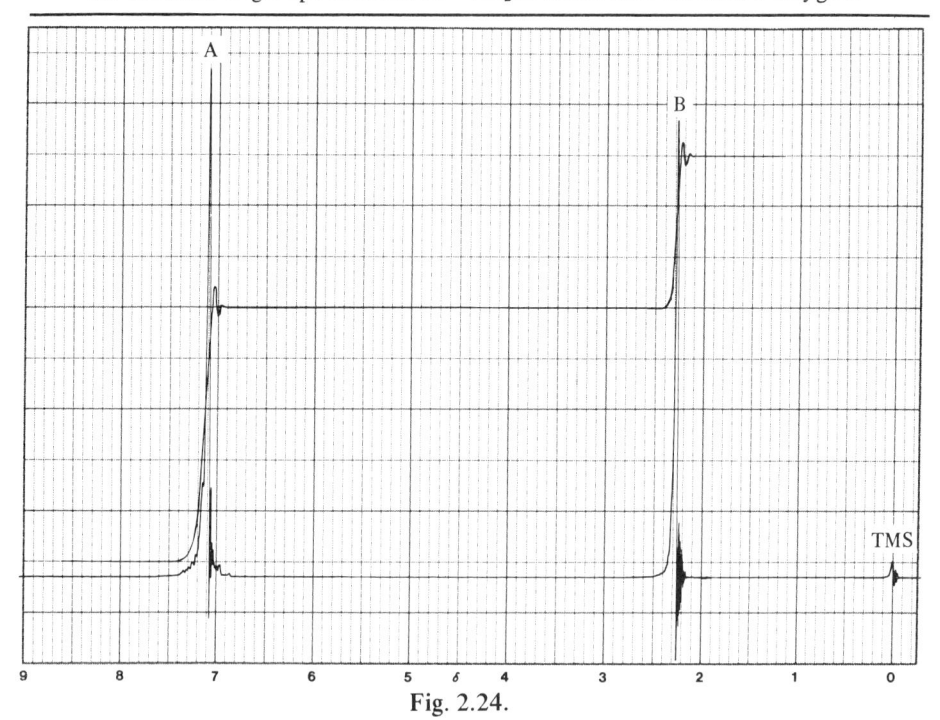

Fig. 2.24.

F91 Figure 2.24 shows the n.m.r. spectrum of a compound C_7H_8. How many protons are associated with each signal?

A90 Yes. Thus structure 22 is consistent with the spectrum in Fig. 2.23.

F92 You have deduced that signal A corresponds to five protons. Classify these protons according to their δ-values.

A91 A, 5 protons; B, 3 protons. The two steps in the integration curve are 34 and 20.5 mm so the 8 protons correspond to 54.5 mm. Thus one proton corresponds to 6.8 mm and the two steps to 5 and 3 protons respectively.

F93 Thus it appears that we are examining the spectrum of either an aromatic compound or a highly conjugated alkene. Taking the former option for the moment and knowing the molecular formula is C_7H_8, write down a possible structure.

A92 Aromatic protons (i.e. protons attached directly to an aromatic ring) or alkenic protons in a conjugated system. The molecular formula rules out heterocycles.

F94 Does signal B have a δ-value consistent with the methyl protons of toluene (26)?

A93

(structure 26: toluene, benzene ring with CH$_3$)

F95 The five aromatic hydrogens of $C_6H_5CH_3$ (27) are not equivalent. This can be seen by comparing the relationship of each proton to the methyl group. There are three different sets of aromatic protons in toluene.

However, their chemical shifts happen to be almost identical in this case and lead to a single absorption as shown in the spectrum. How many chemically different types of aromatic hydrogens are there in each of the following structures (28 to 34)?

(structure 27: toluene with labelled protons H^1, H^1, H^2, H^2, H^3)

(structures 28, 29, 30, 31: substituted toluenes — 28: ortho-dimethylbenzene; 29: 2-bromotoluene; 30: 3-methyl... with CH$_3$; 31: 3-chlorotoluene)

28   29   30   31

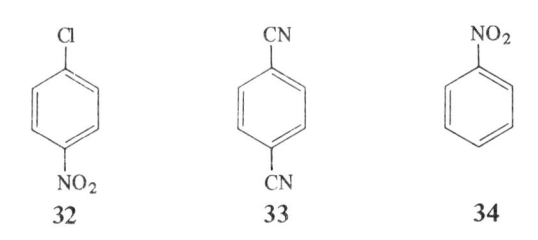

[Structures 32, 33, 34: 4-chloronitrobenzene (Cl / NO₂), 1,4-dicyanobenzene (CN / CN), nitrobenzene (NO₂)]

A94 Yes. You should have found that the B signal (2.3 p.p.m.) lies in the chemical shift range 2.2 to 3.0 p.p.m. given for the $CH_3C_6H_5$ group. The compound is, in fact, toluene as it is impossible to write down a structure of a conjugated alkene which would give this simple n.m.r. spectrum.

F96 The spectrum of an unknown compound $C_{14}H_{12}$ is shown in Fig. 2.25.

How many protons are associated with each signal?

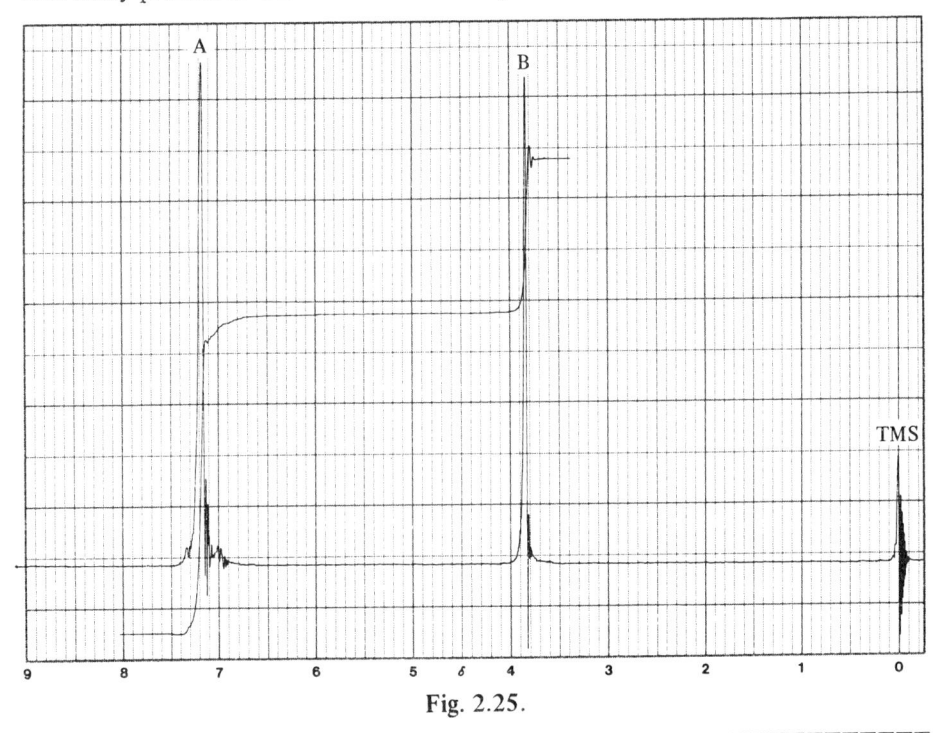

Fig. 2.25.

A95

28—(2): benzene ring with CH₃, CH₃, H¹, H¹, H², H²

29—(4): benzene ring with CH₃, Br, H¹, H², H³, H⁴

30—(3): benzene ring with CH₃, H¹, CH₃, H², H², H³

31—(4): benzene ring with CH₃, H¹, Cl, H², H³, H⁴

32–(2) [structure: chlorobenzene with NO₂ para, showing H¹, H¹, H², H²]

33–(1) [structure: 1,4-dicyanobenzene, four equivalent H]

34–(3) [structure: nitrobenzene showing H¹, H¹, H², H², H³]

F97 Classify the protons responsible for signal A.

A96 Eight for A, four for B.

F98 Take the aromatic option. The compound has fourteen carbon atoms and eight aromatic protons. Assuming the aromatic absorption is due to the presence of benzenoid rings, how many such rings are in the molecule?

A97 Aromatic or alkenic in a conjugated alkene.

F99 We can tentatively conclude that the compound $C_{14}H_{12}$ has eight protons associated with two benzenoid rings. Which of the following structures (**35** to **39**) is consistent with this conclusion?

35 [two benzene rings linked by two CH=CH bridges]

36 [two benzene rings linked by two CH₂ bridges]

37 [two benzene rings linked by CH₂–CH₂]

38 [Ph–CH=CH–Ph]

39 [two benzene rings linked by CH₂–O and O–CH₂]

A98 Two. Eight aromatic protons require two or **more** benzene rings; fourteen carbon atoms require two or **less** benzene rings.

F100 In order to distinguish between **36** and **37** we must turn our attention to the chemical shifts of the non-aromatic protons. The –CH₂– groups of compound **37** are each bonded to one aromatic ring whereas in **36** they are each bonded to two rings. Consult the chart and write down the range in which you would expect to find the absorption position for methylene protons attached to one aromatic ring. Structure **37** has this feature and it *is/is not* consistent with the spectrum given.

A99 **36** and **37**. **35** and **39** do not fit the molecular formula, **38** fits but has ten aromatic protons.

F101 It can be seen from the chart that replacement of an alkyl group by an aromatic ring has a noticeable effect on the δ-values of $-CH_2-$ protons in a saturated aliphatic residue.

$$\text{alkyl}-CH_2-\text{alkyl} \qquad \text{alkyl}-CH_2-\text{aryl}$$
$$\textbf{40} \qquad\qquad\qquad \textbf{41}$$

In fact, in going from formula **40** to **41**, the δ-value is:

(a) raised by approximately 2;
(b) raised by approximately 6;
(c) lowered by approximately 2.

> **A100** is not. The chemical shift range for $RCH_2C_6H_5$ is 2.45 to 3.15 p.p.m. Hence Fig. 2.25 is the spectrum of compound **36**.

F102 So one might expect the protons of a methylene group attached to **two** aromatic rings to have a δ-value somewhat *greater/less* than 2 p.p.m. *above/below* that of methylene protons in a saturated molecule.

> **A101** (a).

F103 You will notice that, as in the spectrum of toluene, the aromatic protons appear as a single absorption. With this in mind examine the structure and decide whether compound **36** has aromatic protons which are:

(a) non-identical, but of very similar chemical shift;
(b) identical;
(c) of widely different chemical shift.

> **A102** greater; above. This comes from extrapolating the deduction in the previous frame.

> **A103** (a).

Shielding and Deshielding Effects of Multiple Bonds

F104 A glance at the chart of δ-values will reveal that protons directly attached to multiple bonds are generally observed at much lower field than those attached to saturated carbon atoms. This is a consequence of the induced field created by the

Fig. 2.26.

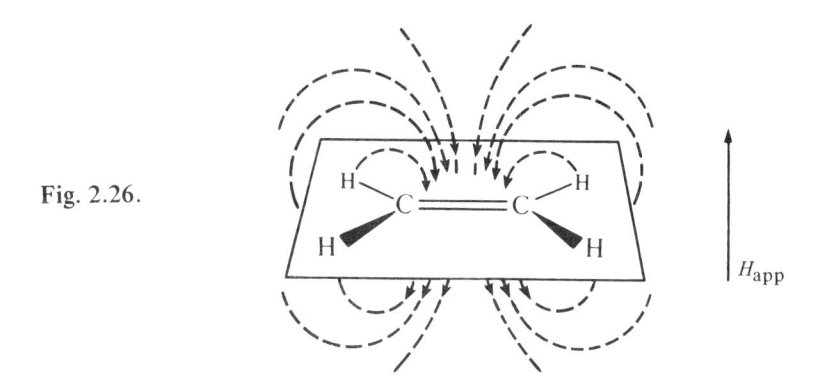

interaction of π-electrons with the external magnetic field, which takes the form shown in Fig. 2.26 for ethylene.

Is the induced field directed with or against the applied field in the neighbourhood of the protons?

F105 Consequently we need *more/less* applied field to reach the resonance condition for these protons than if the double bond had been absent.

> **A104** with. The individual field operates against the applied field between the two carbon atoms.

F106 Consider a proton situated directly above the plane of the molecule at the midpoint of the double bond. Would you expect the n.m.r. signal of such a proton to be observed at higher or lower field than if the double bond was absent?

> **A105** less. The induced field created by the π-electrons is in the same direction as the applied field in the neighbourhood of the proton.

F107 Thus, the space surrounding a double bond can be divided into SHIELDING and DESHIELDING regions in which protons would be observed at relatively high and low fields respectively. The boundary between the regions resembles the surface of a double cone.

Label A and B in Fig. 2.27 as shielding or deshielding regions.

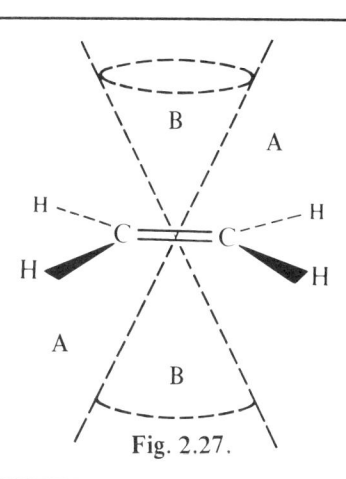

Fig. 2.27.

A106 Higher. See answer 104.

F108 Within region B in Fig. 2.27, the further away a proton is from the double bond the *greater/less* the shielding effect. (Use your intuition!)

A107 A, deshielding; B, shielding.

F109 You should now understand why protons of saturated hydrocarbons are found between 0 and 2 p.p.m. whereas those directly attached to the double bonds of alkenes generally have δ-values greater than 4.5 p.p.m. Which of the following three solvents would be most suitable for the measurement of the n.m.r. spectrum of hexa-1,3,5-triene: methylene chloride, n-hexane, benzene? (CH_2Cl_2, 5.3 p.p.m.)

A108 less.

F110 Predict the n.m.r. spectrum of compound **42** giving the number of signals, their relative areas and approximate δ-values.

42

A109 n-Hexane, because the chemical shifts of saturated hydrocarbons are well outside the range of shifts expected for alkenic protons.

F111 The carbonyl group is another multiple bond commonly found in organic compounds, and the electronegativity difference between carbon and oxygen leads to the bond being polarized in the sense shown (A).

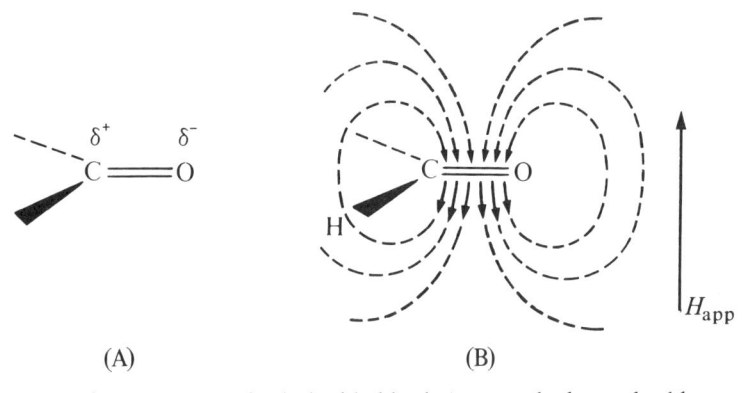

(A) (B)

Aldehydic protons, therefore, are not only deshielded by being attached to a double bond (B), but also by the inductive pull of the electron deficient carbon atom on the electrons of the C–H bond. Consequently, it is not surprising to find δ-values of aldehydic protons corresponding to very low fields (about 10 p.p.m.). Which of the following compounds would be expected to have a signal close to 10.0 p.p.m.?

$$\begin{matrix} CH_3 \\ Cl \end{matrix} C=O \qquad HCHO \qquad \begin{matrix} CH_2=CH \\ CH_3 \end{matrix} C=O \qquad \text{cyclopentyl} \begin{matrix} CHO \\ CH_3 \end{matrix}$$

43　　　　44　　　　　45　　　　　　46

A110

Compound 47: cyclopentene-like structure with CH_3^a groups, H^c protons on the double bond carbons, and H_2^b.

Three signals:

a – 1 p.p.m. area 6
b – 1.5 p.p.m. area 1
c – 6 p.p.m. area 1

F112 The electronegativities of carbon, nitrogen and oxygen increase in that order. Which of the δ-values, 11.0, 8.0 and 5.0 p.p.m. is most appropriate for the signal of a proton attached to an imine group, e.g.

$$\begin{matrix} R^1 \\ H \end{matrix} C=N_{R^2} \qquad R^1 \text{ and } R^2 = \text{alkyl}.$$

48

A111 44 and 46.

F113 It is well known that benzene does not normally behave as a cyclic triene, but is endowed with a special character termed AROMATICITY. The aromatic sextet of electrons present in benzenoid compounds constitutes a stable 'closed shell' configuration and also leads to unique magnetic properties. Thus the induced field created by the interaction of benzenoid π-electrons with an applied magnetic field, can be thought of as resulting from a RING CURRENT as shown in Fig. 2.28.

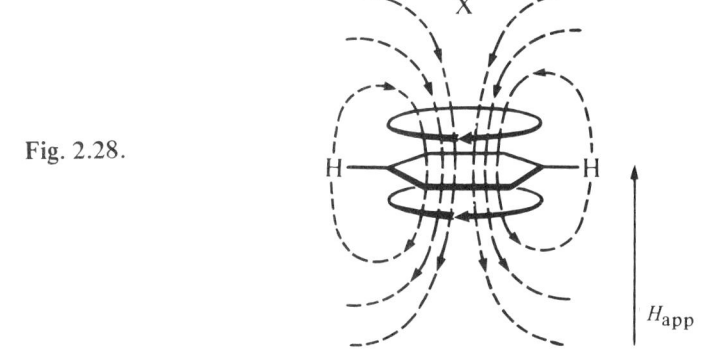

Fig. 2.28.

Are the protons drawn in Fig. 2.28 heavily shielded or deshielded?

A112 8.0 p.p.m. The electronegativity of nitrogen lies between that of carbon and oxygen.

F114 Would a proton directly above or below the ring (e.g. at point X in Fig. 2.28) resonate at exceptionally high or low field?

A113 Deshielded. The induced field created by the π-electrons of the benzene ring is in the same direction as the applied field in the neighbourhood of the proton.

F115 Thus, as in the case of alkenes, the space surrounding a benzenoid ring can be divided into shielding and deshielding regions. Which is which in Fig. 2.29?

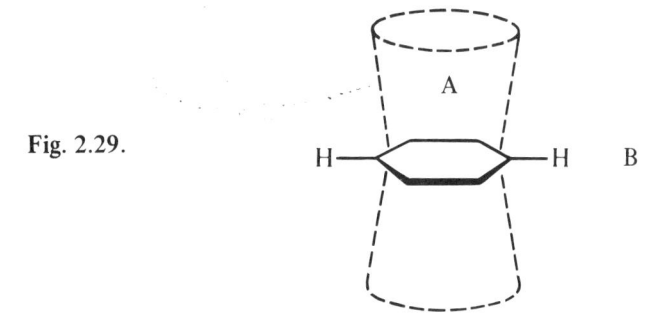

Fig. 2.29.

A114 High field. The induced field created by the π-electrons operates against the applied field in the region directly above and below the plane of the ring.

F116 Decamethylenebenzene (**49**) is an example of a compound with protons lying in both the shielding and deshielding regions of the benzenoid ring. Recalling that methylene protons usually absorb in the 1.2 to 1.7 p.p.m. region, select the most appropriate pair of δ-values for the signals of the protons at A and B.

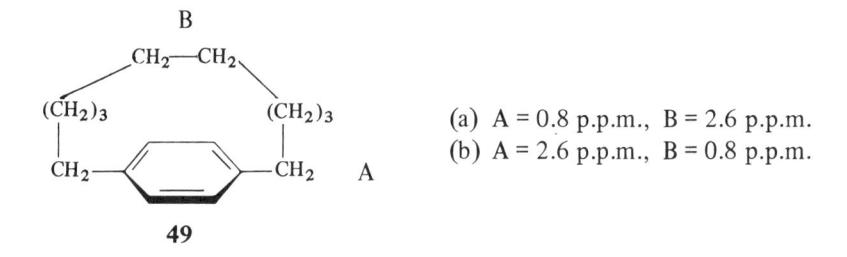

(a) A = 0.8 p.p.m., B = 2.6 p.p.m.
(b) A = 2.6 p.p.m., B = 0.8 p.p.m.

A115 A, shielding; B, deshielding.

F117 The shielding and deshielding effects of benzenoid ring currents are more powerful than the deshielding and shielding properties of the π-electrons in alkenic bonds. Thus, whereas protons attached to isolated double bonds are observed to resonate between 4.6 and 6.4 p.p.m. benzenoid protons usually resonate between 6.6 and 8.5 p.p.m. Would you expect protons *a* in **50** to resonate at higher or lower field than protons *b* in **51**?

50

51

A116 A = 2.6 p.p.m. (in the deshielding region), B = 0.8 p.p.m. (in the shielding region).

F118 If an unknown compound shows an n.m.r. signal with a δ-value higher than 6.6 p.p.m. there is a strong likelihood that the proton(s) responsible is(are) deshielded by a ring current. It has been established theoretically and experimentally, however, that ring currents are not peculiar to benzenoid compounds alone but are a characteristic of all aromatic species. Thus, the large macrocycle 18-annulene (**52**) was judged to be aromatic on the basis of n.m.r. evidence of a ring current. Two signals are observed with area ratio 1:2. The δ-values were 8.3 and −1.8 p.p.m.; which δ-value corresponds to the smaller signal?

52

A117 Higher. H^a and H^b are in the shielding regions of the aromatic ring and the double bond respectively. The shielding effects of the benzenoid ring is stronger than that of the alkenic double bond.

F119 N.m.r. evidence is now widely used as a criterion for aromaticity. For example, does the fact that the methyl protons of 15,16-dihydro-15,16-dimethylpyrene (**53**) absorb at −4.2 p.p.m. support the claim that the compound is aromatic? Explain.

53

A118 −1.8 p.p.m. The six inner protons lie in the shielding region within the ring.

F120 By no means all signals having δ-values greater than 6.6 p.p.m. are associated with 'aromatic' protons. Give an example of a type of proton discussed recently which resonates at low field but which is not deshielded by a ring current. (You should be able to do this without looking at the chart!)

A119 Yes. The negative δ-value implies that the methyl groups lie in the shielding region of a ring current.

F121 Whilst considering multiple bonded systems we have so far confined our attention to double bonds. Finally, let us consider the resonance position of acetylenic protons. Reference to the chart tells us that they are *more/less* shielded than alkenic protons.

A120 An aldehydic proton, $\overset{\displaystyle\diagdown}{\underset{H}{\diagup}}C=O$

F122 In the presence of an applied field the electrons of the cylindrical π-cloud give rise to an induced field as shown in Fig. 2.30, so that acetylenic protons need a *greater/smaller* applied field to reach the resonance condition than if this effect had been absent.

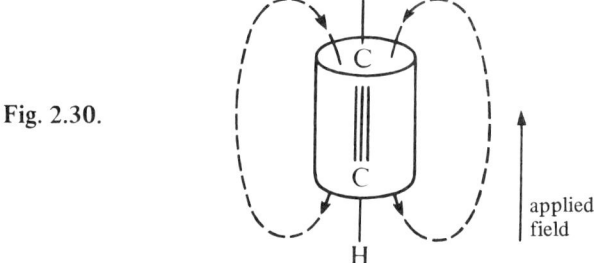

Fig. 2.30.

A121 more. Acetylenic protons resonate at higher field strength.

F123 Which of the following statements is/are true?

(a) Aromatic protons are more heavily shielded than acetylenic protons.
(b) Protons attached to a benzene ring nearly always have δ-values greater than 6.6 p.p.m.
(c) All multiple bonds give rise to ring currents.

A122 greater. The field induced by the π-electrons operates against the applied field in the neighbourhood of the acetylenic protons.

F124 We have seen that the magnetic environment of a multiple bond is ANISOTROPIC, that is, it is not uniform in all directions, but is critically dependent on the geometric relationship of the position concerned to the bond itself. In other words multiple bonds give rise to induced fields which are ANISOTROPIC, and compounds containing multiple bonds are referred to as being DIAMAGNETICALLY ANISOTROPIC. In general it is advisable to avoid this property when selecting a solvent for n.m.r. measurement, because the chemical shift of the sample will depend on the detailed nature of solvation, and this will vary for each compound measured. Which of the following solvents are diamagnetically anisotropic?

C_6D_6 (hexadeuterobenzene)
54

$CDCl_3$
55

CF_3CO_2H
56

CD_3COCD_3
57

$Cl_2C=CCl_2$
58

CCl_4
59

nitrobenzene
60

A123 (b). (a) is false because observed δ-values for aromatic protons are greater than those for acetylenic protons; (c) is false because we have stated that ring currents as defined are characteristic of aromatic species only.

A124 All except $CDCl_3$ and CCl_4.

Revision Summary

Shielding

Single bonded systems

In practice we do not encounter isolated protons in organic chemistry, but compounds in which the hydrogen nuclei are surrounded by valence electrons. The motion of these electrons, induced by an applied magnetic field, gives rise to a small secondary field in the vicinity of the proton. This induced field operates against the applied field, and shields the proton from its influence. In order to bring the proton into resonance, the applied field must be increased to overcome the DIAMAGNETIC SHIELDING effect, i.e. for resonance:

$$H = H_0 + H'$$

normal applied field — extra applied field applied to overcome shielding effect

The magnitude of the shielding depends largely upon the electron density in the region of the proton and consequently protons in different chemical environments are magnetically different and need different field increments to bring them into resonance.

The three sets of protons in CH_3CH_2OH resonate at different field strengths (see diagram).

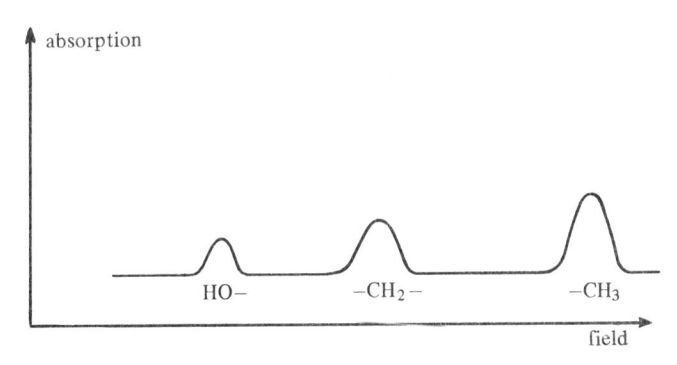

[Graph: absorption vs field, showing three peaks labeled HO–, –CH₂–, –CH₃ from left to right]

The electronegativity of the neighbouring atoms (the inductive effect) plays a large part in determining the electron density in the vicinity of the proton. Consequently, the greater the inductive effect to which the proton is subjected, the lower the extra field required to bring about resonance.

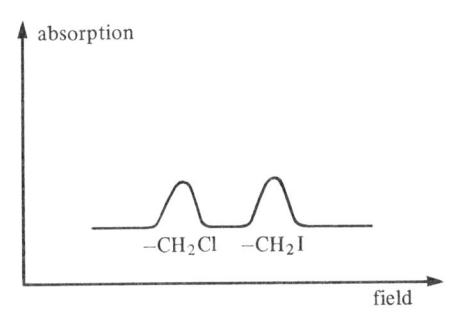

[Graph: absorption vs field, showing two peaks labeled –CH₂Cl and –CH₂I]

For $ClCH_2CH_2I$, the methylene protons attached to iodine absorb at higher field than those attached to the more electronegative chlorine.

Shielding in multiple-bonded systems
Multiple bonds give rise to induced magnetic fields which divide the immediate environment into two regions; one in which a proton is shielded, the other in which it is deshielded from the applied field.

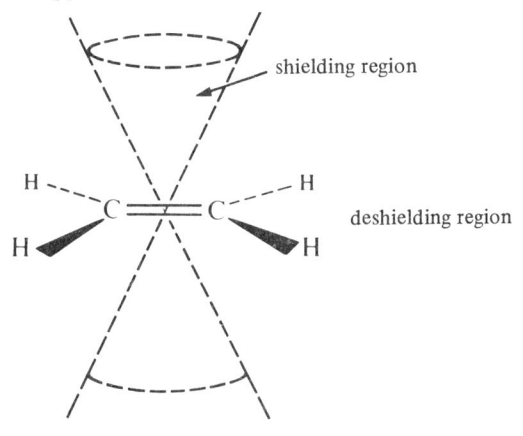

[Diagram of C=C double bond with H substituents, showing shielding region (cones above and below) and deshielding region]

(a) *Alkenes.* Alkenic protons are deshielded and hence have greater δ-values (4.6 to 7.7 p.p.m.) than their saturated counterparts.

(b) *Aldehydes.* The inductive effect of the carbonyl group increases the deshielding effect, hence aldehydic protons resonate at particularly low fields (9.3 to 10.0 p.p.m.).

(c) *Aromatic compounds.* The π-electrons are particularly susceptible to an applied field and a relatively large (compared with alkenes) induced field is set up. Consequently benzenoid protons (6.6 to 8.5 p.p.m.) and aromatic protons in general have greater δ-values than alkenic protons. (This is the basis of the use of n.m.r. spectroscopy in recognizing aromaticity.)

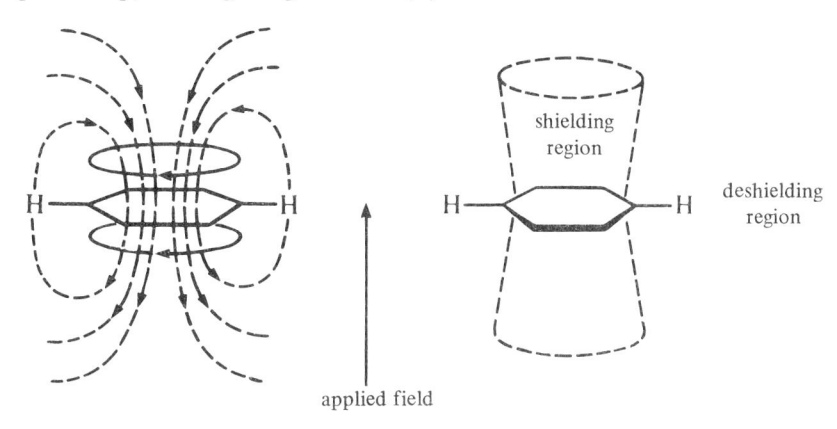

(d) *Alkynes.* Acetylenic protons lie in a shielding region and hence have lower δ-values (1.8 to 3.1 p.p.m.) than alkenic protons.

Integration

The area under the absorption band is directly proportional to the number of protons responsible for the signal. It is measured electronically and superimposed on the spectrum as an INTEGRATION CURVE.

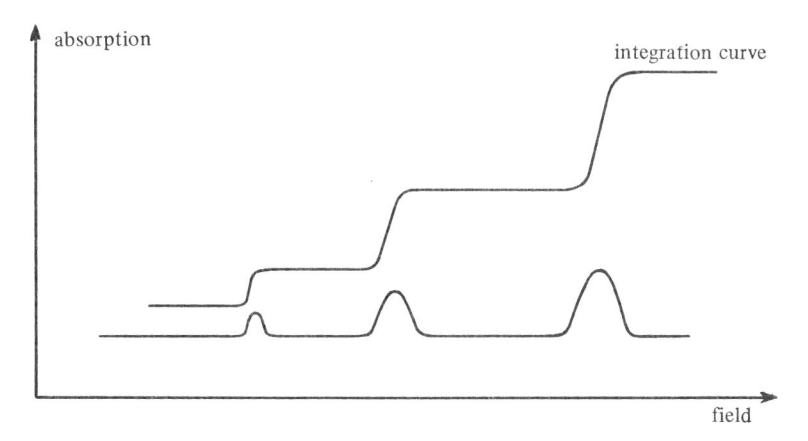

The ratio of the height of the steps in the curve gives the ratio of the number of different protons associated with each signal.

Chemical shift

The positions of the absorption bands in the spectrum are recorded quantitatively with respect to a standard (tetramethylsilane – TMS). The difference between the resonance position of a particular proton and that of the TMS protons is known as the CHEMICAL SHIFT and is recorded as:

$$\delta \text{ (parts per million)} = \frac{\text{frequency of TMS signal} - \text{frequency of sample signal}}{\text{spectrometer frequency}} \times 10^6$$

Alternatively, the tau-value ($\tau = 10 - \delta$) is recorded. Both δ- and τ-values are characteristic of the particular type of proton and are independent of the instrument used.

Correlation charts

There is a close correlation between the chemical shift and the chemical environment of the proton. Consequently correlation charts have been produced (e.g. Chart 1) which can be used in an empirical manner to assist in the interpretation of a spectrum.

The positions of protons bound to oxygen and nitrogen are not included on Chart 1 as their chemical shift values vary considerably depending upon the concentration of the solution, the nature of the solvent and the temperature.

Now try the multiple choice test at the beginning of this part. Write your answers in a vertical column and mark your script as before. Then consult the remarks below.

0–6: Consult your instructor as to whether you should go on or seek extra tuition. If you are using the programme for private study, re-read the frames pertinent to those you answered incorrectly and, when you are satisfied you fully understand the material, read the paragraph at the end of this part and start Part 3.

More than 6: Revise the material pertinent to those questions you answered incorrectly. If your instructor does not want you to consult the bibliography or attempt any of the questions which follow, read the paragraph at the end of this part and continue with Part 3.

Questions

1. Suggest possible substituents for the furan derivative $C_7H_8O_3$ whose n.m.r. spectrum is shown below. Furan (**61**), has some aromatic character.

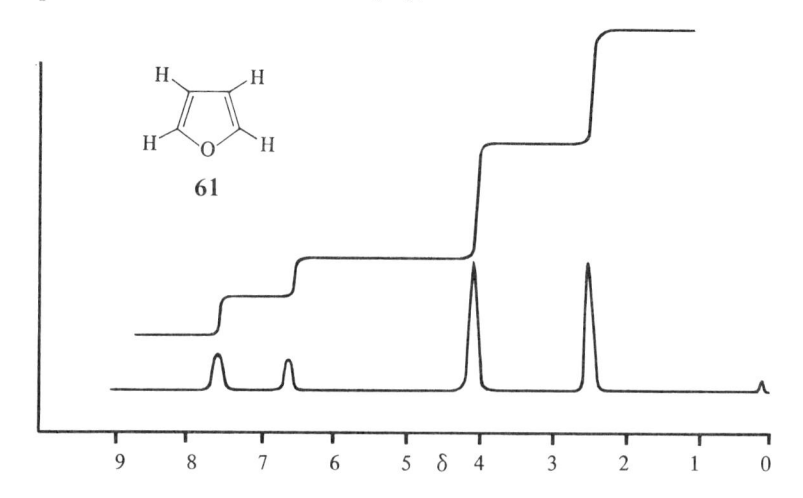

2. Deduce the structure of the 1,1-disubstituted ethylene C_5H_6 whose low resolution n.m.r. spectrum is given below.

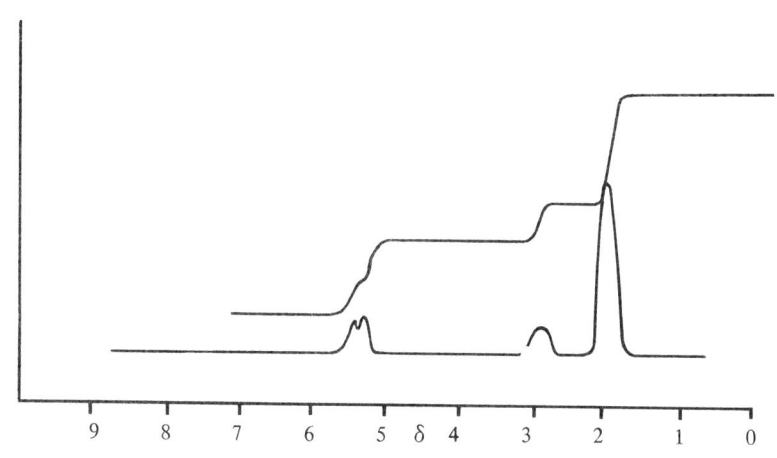

3. The low resolution n.m.r. spectrum of compound **62** has five absorption bands at 1.8, 1.9, 3.8, 4.5 and 6.2 p.p.m. Assign these to the different protons *a* to *e* in the structure.

62

4. Which of structures **63**, **64** and **65** is consistent with the spectrum below?

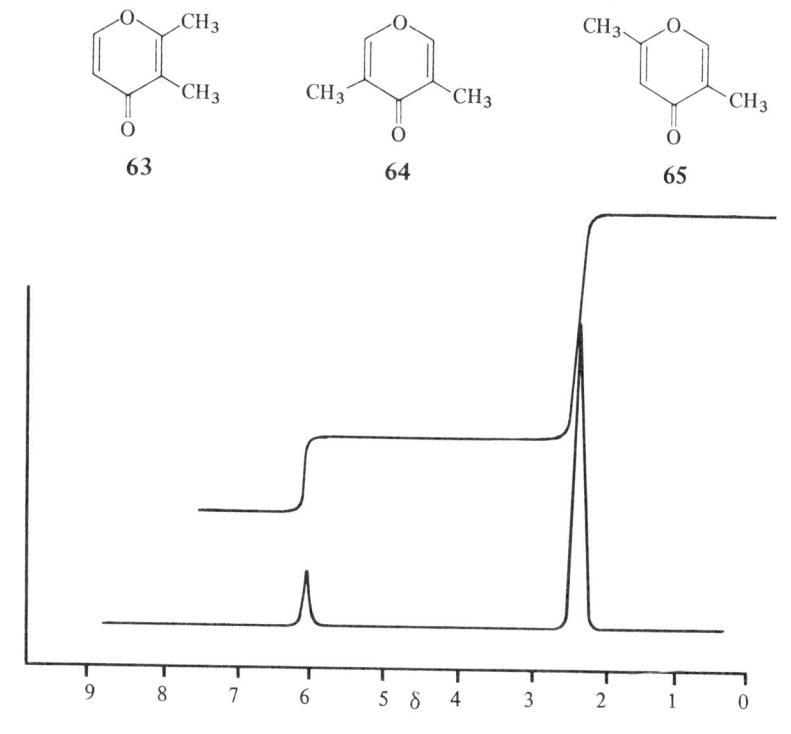

63

64

65

5. A student was asked to reduce a sample of benzaldehyde to benzyl alcohol.

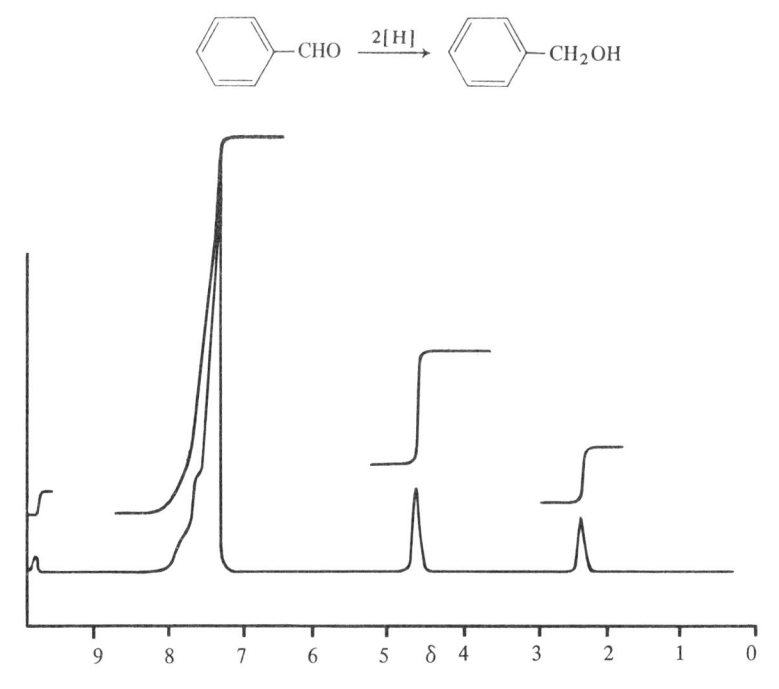

$$\text{C}_6\text{H}_5\text{-CHO} \xrightarrow{2[H]} \text{C}_6\text{H}_5\text{-CH}_2\text{OH}$$

After a short time he removed a portion from the reaction mixture, isolated the organic components, dissolved them in carbon tetrachloride containing a trace of TMS and the solution gave the n.m.r. spectrum shown above. How far had the reaction proceeded?

Further Reading

L. M. Jackman and S. Sternhell, *Applications of Nuclear Magnetic Resonance in Organic Chemistry*, 2nd Edition, Pergamon Press, Oxford, 1969, Chapters 2.2 and 3.1 to 3.9 inclusive.

D. W. Mathieson, *Nuclear Magnetic Resonance for Organic Chemists*, Academic Press, London, 1967, Chapters 2 and 3.

J. W. Emsley, J. Feeney and L. H. Sutcliffe, *High Resolution Nuclear Magnetic Resonance Spectroscopy*, Volume 1, Pergamon Press, Oxford, 1965, Chapter 3, paragraphs 1 and 3 to 12 inclusive.

D. Chapman and P. D. Magnus, *Introduction to Practical High Resolution Spectroscopy*, Academic Press, London, 1966.

J. R. Dyer, *Applications of Absorption Spectroscopy of Organic Compounds*, Prentice-Hall, New Hersey, 1965, Chapter 4, paragraphs 3 and 7.

R. M. Silverstein and G. C. Bassler, *Spectrometric Identification of Organic Compounds*, John Wiley & Sons, New York, 1967, Chapter 4, paragraph 3.

D. H. Williams and I. Fleming, *Spectroscopic Methods in Organic Chemistry*, McGraw-Hill, Maidenhead, 1966, Chapter 4, paragraphs 3 and 4.

R. H. Bible, Jr., *Interpretation of N.m.r. Spectra*, Plenum Press, New York, 1965, p. 7.

S. F. Dyke, A. J. Floyd, M. Sainsbury and R. Theobald, *Organic Spectroscopy – An Introduction,* Penguin, 1971, Chapter 4, section 4.4.

E. D. Becker, *High Resolution N.m.r.,* Academic Press, New York, 1970, Chapter 4.

F. A. Bovey, *Nuclear Magnetic Resonance Spectroscopy,* Academic Press, New York, 1967, Chapter 3.

W. W. Paudler, *Nuclear Magnetic Resonance,* Allyn and Bacon, Boston, 1971, Chapter 2.

You are now able to correlate what we have called low resolution spectra, with molecular structure, and solve straightforward structural problems by n.m.r. spectroscopy.

You know how to make use of empirical information and are able to recognize different types of protons with the aid of a correlation chart. You have learnt how to rationalize the relative δ-values of different protons given in the chart on the basis of the theory of shielding and can use the theory to predict the δ-values of those protons whose precise environment is not covered by the chart. You have seen that shielding is modified by the inductive effect and the presence of multiple bonds in the molecule. Finally, you appreciate that the area of each absorption band is directly proportional to the number of associated protons and you have learnt how this important information can be quickly abstracted from the spectra by an examination of the superimposed integration curve.

PART 3
SPIN—SPIN COUPLING

Aim
After completing this part of the programme, the student will be able to extract additional molecular structural information about a compound from its n.m.r. spectrum by analysis of first-order multiplet structures of signals and by the use of coupling constants. He will have acquired experience in the interpretation of n.m.r. spectra having been guided in the combined use of integration, chemical shift, and spin—spin coupling data.

New Terms and Concepts
Spin—spin coupling, coupling constant.

Prior Knowledge
Parts 1 and 2; the structures of simple aromatic and aliphatic compounds.

Objectives for Part 3

When you have completed this part you should be able to:

1. Use first-order n.m.r. coupling patterns to select the correct structure of an unknown from a given set of alternatives.
2. Given the structure of an organic compound, predict the shapes of n.m.r. signals involved in first-order coupling.
3. Identify the features in coupling patterns which constitute measures of coupling constants.
4. Deduce relative sizes of coupling constants from signals of protons coupled to two other different sets of protons.
5. Use a table of coupling constant values to predict the shape of a signal of a proton coupled to more than one set of identical protons.
6. Use the principles of spin–spin coupling to interpret the signals of O–H protons, N–H protons and C–H protons adjacent to O–H and N–H bonds.
7. Predict the ratios of the components of any given first-order multiplet.
8. Use correlation charts of chemical shifts and coupling constants to deduce the structure of an unknown organic compound of given molecular formula and n.m.r. spectrum. (The level of difficulty intended in this objective is best gauged by referring to Figs. 3.19 to 3.21.)

Introduction

Most n.m.r. spectra can be described as a series of absorption bands (distributed along a field scale) many of which are 'split' into two or more components. We showed in Part 2 the way in which valuable information can be extracted from the field positions of these bands and their relative areas. Part 3 is designed to give an understanding of the reasons why the bands are 'split' into components, and to enable you to analyse band structures to an extent which leads to additional information about the material being examined.

Attempt the following multiple choice test, marking it in the usual way by reference to the correct answer sequence on p. 175 and then compare your marks with the comments set out after the test. Do not seek the correct answers to incorrectly answered questions at that stage.

Multiple Choice Test

1. The coupling pattern in (Fig. 3.1) is produced by:

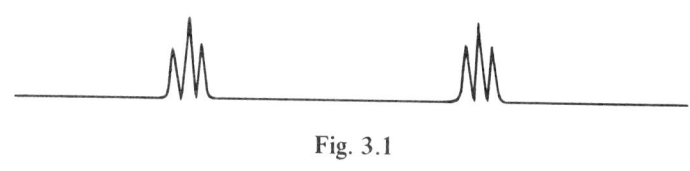

    Fig. 3.1

    (a) Y—CH$_2$—CH$_2$—X,   (b) $\begin{array}{c}Y\\Y\end{array}$CH—CH$\begin{array}{c}X\\X,\end{array}$ or   (c) CH$_3$—C(X)(Y)—CH$_3$

    where X and Y are groups of widely differing electronegativities.

2. The signal for H^a in compound 1 is drawn in Fig. 3.2.

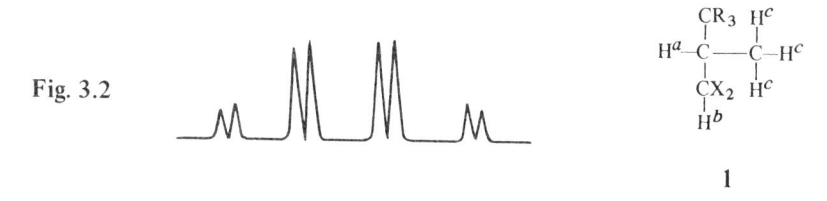

    Fig. 3.2

    Compound 1: H^a—C(CR$_3$)(CX$_2$)—C(H^c)(H^c)—H^c, with H^b on the CX$_2$ carbon.

    Which of the following conclusions is correct?
    (a) $J_{ab} > J_{ac}$,   (b) $J_{ab} = J_{ac}$,   (c) $J_{ab} < J_{ac}$

3. Which of the following statements is **false**?
    (a) Hydroxylic protons may be observed as multiplets if the rate of proton exchange is sufficiently reduced.
    (b) NH protons do not normally appear as well-defined multiplets.
    (c) Unlike those of OH protons, the signal shapes of NH protons are independent of conditions in the n.m.r. sample tube.

4. The coupling constant J in the triplet (Fig. 3.3) is defined as the separation (a) x, (b) y, (c) $y/2$.

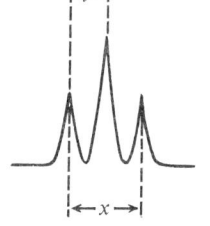

Fig. 3.3

5. Which of the absorption patterns in Fig. 3.4 is **least** likely to have arisen from the fragment $-NHCH_2CCl_3$?

(a)   (b)   (c)

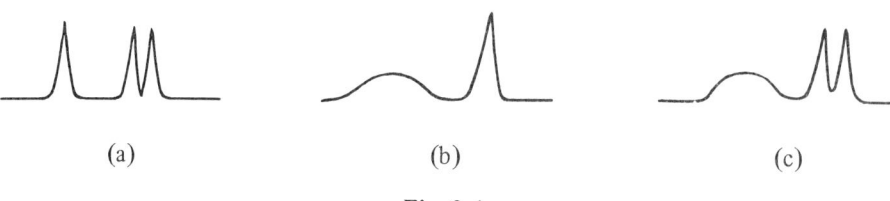

Fig. 3.4

6. Which of compounds **2, 3** and **4** could give the partial spectrum in Fig. 3.5?

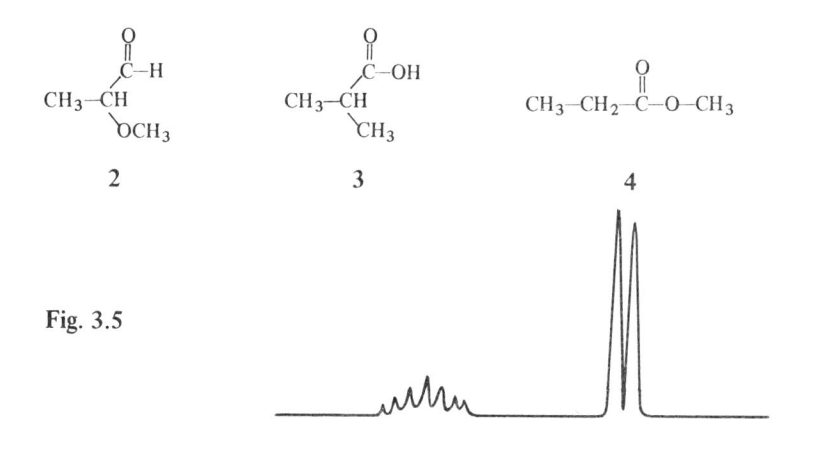

$$\begin{array}{ccc}
CH_3-\underset{\underset{OCH_3}{|}}{CH}-\overset{O}{\underset{\|}{C}}-H & CH_3-\underset{\underset{CH_3}{|}}{CH}-\overset{O}{\underset{\|}{C}}-OH & CH_3-CH_2-\overset{O}{\underset{\|}{C}}-O-CH_3 \\
\mathbf{2} & \mathbf{3} & \mathbf{4}
\end{array}$$

Fig. 3.5

7. How many methylene groups would appear as singlets in the n.m.r. spectrum of compound **5**?
(a) 1, (b) 3, (c) 7.

5

8. Which of compounds **6, 7** and **8** would you expect to give an n.m.r. spectrum containing both a simple triplet and a simple doublet?

$$\begin{array}{ccc}
CH_3-O-CH_2-CO-CH_2-OCH_3 & CH_3-CO-O-CH_2-\underset{\underset{CHO}{|}}{\overset{\overset{OCH_3}{|}}{CH}} & CH_3-CO-CH_2-\underset{\underset{OCH_3}{|}}{\overset{\overset{OCH_3}{|}}{CH}} \\
\mathbf{6} & \mathbf{7} & \mathbf{8}
\end{array}$$

9. The methine proton (CH) of 3-cyanopentane (**9**) is expected to have an n.m.r. signal with five components in the ratio:

   (a) 1:5:8:5:1
   (b) 1:3:6:3:1
   (c) 1:4:6:4:1

$$CH_3-CH_2-\underset{\underset{CN}{|}}{\overset{\overset{H}{|}}{C}}-CH_2-CH_3$$

9

10. Consult Chart 2 and deduce whether the methylene signal in the n.m.r. spectrum of the alkene **10** will appear as:

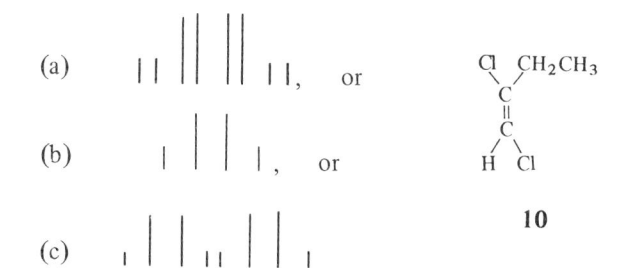

   (a), (b), or (c)

$$\underset{HCl}{\overset{ClCH_2CH_3}{\underset{\|}{\overset{\diagdown\diagup}{C}}\underset{\diagup\diagdown}{C}}}$$

10

Comments on marks

Less than 2: You need instruction — continue with Part 3.

3–7: You have some knowledge of this topic but it would be in your interest to secure a more thorough understanding and you are advised to work through Part 3 — it shouldn't take you long!

More than 7: Unless you obtained full marks, consult the sections of Part 3 relevant to your incorrect answers; then proceed directly to Part 4.

PART 3: SPIN–SPIN COUPLING

The Origin of Spin–Spin Coupling

F125 We stated in the introduction to Part 2 that in order to focus attention on chemical shift, we would temporarily ignore a second important feature of n.m.r. spectroscopy, namely, spin–spin coupling. Whenever possible the examples chosen were of compounds whose n.m.r. spectra did not show spin–spin splitting. In the few instances when this was not feasible, the characteristics of spin–spin coupling were concealed by the use of 'low resolution' spectra. For example the spectrum of ethanol (frame 43) was shown as consisting of three peaks attributed to the methyl, methylene and hydroxyl protons. Under high resolution, however, the spectrum (Fig. 3.6) shows that two of the signals are multiplets.

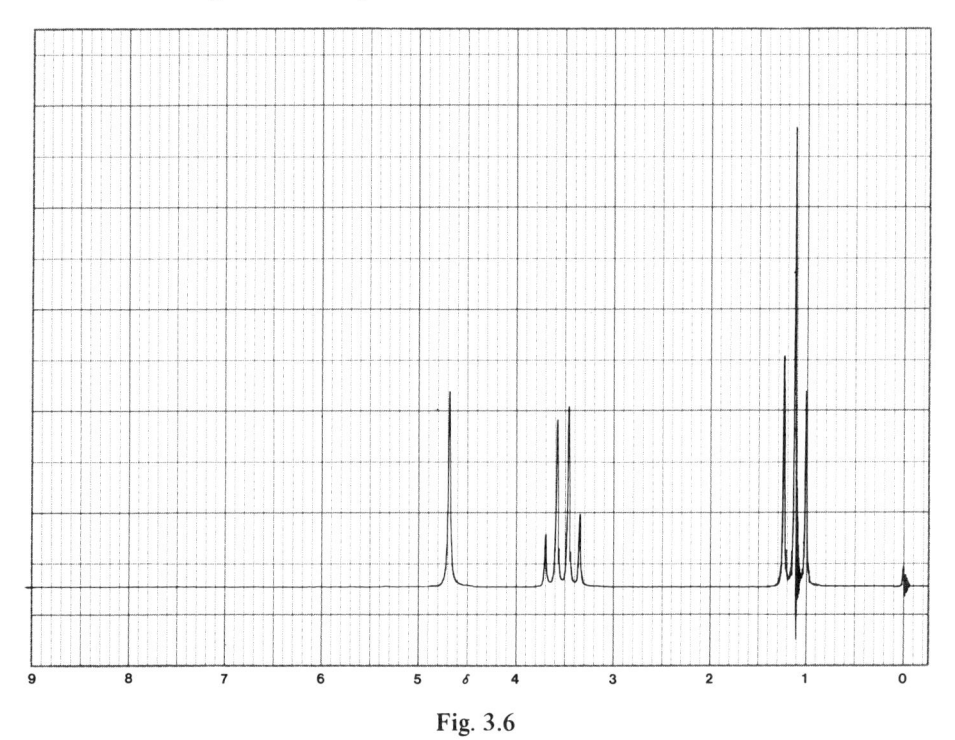

Fig. 3.6

Into how many peaks have (a) the methyl, and (b) the methylene signals been resolved?

F126 We shall begin, however, by examining a very simple case. Consider the signal of a proton H^a in compound **11**. You have learnt that in the recording of its nuclear magnetic resonance signal, H^a will come under the influence of two magnetic fields — the applied field H_0 and a local induced field created by the electrons in the molecule (H'). In practice, however, it is also influenced by the proton H^b. This influence can either add to or subtract from the field required for resonance ($H_0 + H'$), depending on which orientation H^b assumes with respect to the applied field. We stated in Part 1 that the SPIN QUANTUM NUMBER (I) limited the actual number of orientations to $2I + 1$. How many components will be observed in the H^a signal?

> **A125** (a) Three (a triplet), (b) four (a quartet). Recall or see from Chart 1 the relative chemical shifts of CH_3CR_2X and RCH_2O-.

F127 There is an almost equal probability at any one time of H^b lying with or against the applied field. On this basis, would you expect the n.m.r. signal of H^a to appear as (a), (b) or (c) in Fig. 3.7?

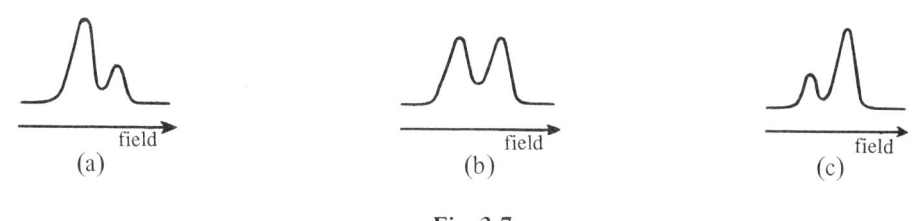

Fig. 3.7

> **A126** Two. The adjacent proton H^b can only assume two orientations in the applied field as the spin quantum number of a proton is ½;

F128 Such an effect is mutual on the protons taking part. Will the spectrum of the compound shown in frame 126 (11) consist of (a) a doublet and a singlet, (b) two doublets, (c) two singlets?

> **A127** (b). The influence of H^b can either add to or subtract from the applied field and there is an almost equal probability of both situations existing.

F129 The two protons are said to be coupled and the phenomenon is known as spin–spin coupling. The extent to which the signal is split is the COUPLING CONSTANT, normally given the symbol J and measured in hertz. Now the magnitude of J is determined only by the nature of the link between the protons involved. Would you say, then, that (a) $J_{ab} = J_{ba}$, (b) $J_{ab} \neq J_{ba}$ or (c) J_{ab} sometimes equals J_{ba} depending on X and Y?

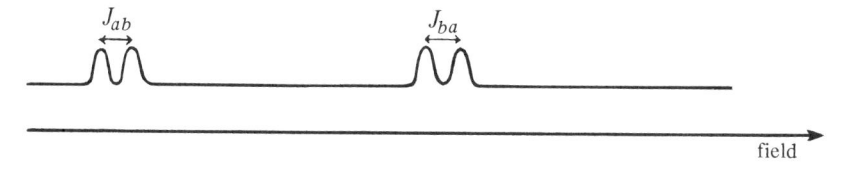

Fig. 3.8

> **A128** (b) Two doublets. The shape of the H^b signal can be deduced using the same arguments as those used for the H^a signal.

F130 Which of the following statements is/are true?
(a) J is independent of the applied field H_0.
(b) Coupling constants are measured in hertz.
(c) Spin–spin coupling simplifies the appearance of n.m.r. spectra.

A129 (a) $J_{ab} = J_{ba}$. The nature of the link between the protons H^a and H^b is always the same as that between H^b and H^a!

F131 It appears that coupling between protons is relayed through bonds and not directly through space. It falls off rapidly with distance and is normally only significant in saturated systems when protons are separated by three or fewer bonds.

$$\begin{array}{c} H^a \quad CH_3 \\ | \quad | \\ CH_3-C-C-H \\ | \quad | \\ Cl \quad COCH_3 \end{array}$$

12

How many protons may be noticeably coupled to H^a in the molecule shown in structure **12**?

A130 (a) and (b). J (in hertz) is only dependent on the nature of the link between the protons involved. (c) is false as spin–spin coupling causes signals to occur as multiplets.

F132 We have considered the simple case of two magnetically different protons coupled to one another giving rise to two doublets. Now let us turn to the situation where one proton, H^a is coupled to two identical protons, H^b, as in chloro-acetaldehyde (**13**). The field required to induce H^a to absorb radiation will be made up from the applied field, the field due to the electrons in the molecule and, at any one instant, the influence of the orientations of the two protons H^b. The first two factors determine the chemical shift of H^a. The influence of H^a on the two protons H^b will depend on the latter's orientation with respect to the applied field. Each H^b proton can have its magnetic moment vector lying with or against the field and so this leads to four possible combinations:

$$\begin{array}{c} H^a \\ \diagdown \\ O=C-CH_2^b-Cl \end{array}$$

13

$$\begin{array}{cccc} \quad H\uparrow & \quad H\downarrow & \quad H\uparrow & \quad H\downarrow \\ OHC-C-Cl & OHC-C-Cl & OHC-C-Cl & OHC-C-Cl \\ \diagdown H\uparrow & \diagdown H\downarrow & \diagdown H\downarrow & \diagdown H\uparrow \\ (a) & (b) & (c) & (d) \end{array}$$

$\uparrow H_0$

Two of these combinations are indistinguishable with respect to the applied field. Which are they?

A131 Four, as shown in **14**.

$$\begin{array}{c} H \quad H^a \\ | \quad | \\ H-C-C-C-H \\ | \quad | \quad | \\ H \end{array}$$

14

F133 There is an almost equal probability of a proton assuming each orientation and thus combinations (a), (b), (c) and (d) are equally probable. You have deduced that (c) and (d) are equivalent. Will H^a appear as (i) a triplet of area ratio 1:3:1, or (ii) a quartet of area ratio 1:1:1:1, or (iii) a triplet of area ratio 1:2:1?

A132 (c) and (d). Each has one magnetic moment vector aligned with and one against the applied field.

F134 Now consider the signal due to H^b. These protons are equivalent and so they will have the same chemical shift, i.e. they will absorb at the same position. The precise position at any one instant will be determined by the orientation of H^a with respect to the applied field.

Thus the protons H^b will be observed as (a) a singlet, (b) a doublet, (c) a triplet.

A133 (iii) A triplet of area ratio 1:2:1.

$$(\downarrow\downarrow \ : \ \downarrow\uparrow \ \uparrow\downarrow, \ : \ \uparrow\uparrow)$$

F135 Which of the spectra in Fig. 3.9 resembles that expected for chloroacetaldehyde?

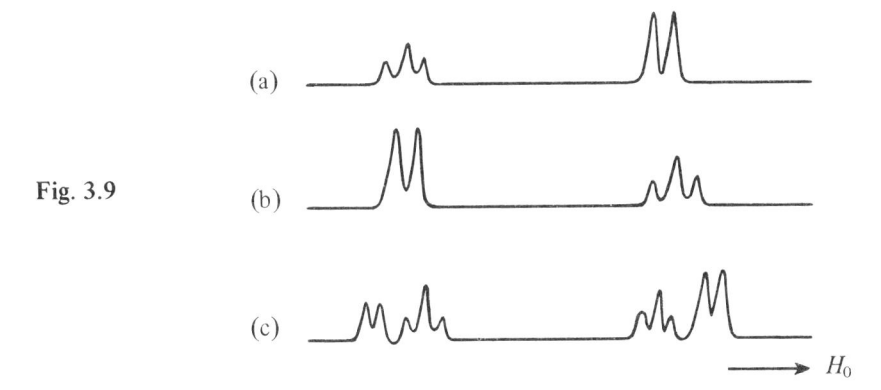

Fig. 3.9

(a), (b), (c)

$\longrightarrow H_0$

A134 (b) a doublet. The two equivalent protons, H^b, may be regarded for this purpose as a single absorbing system whose resonance position can have one of two values depending upon the two orientations which can be adopted by H^a.

F136 Let us now consider the spectrum of ethyl acetate (15), which contains three sets of magnetically different protons. The three protons H^a are all equivalent and five bonds removed from their nearest neighbours H^b. We can confidently predict that spin–spin coupling between H^a and H^b will be *significant/insignificant*.

15

74

A135 (a). (c) is inconsistent with the description of the two signals given in the previous frames. (b) shows the aldehydic proton absorbing at higher field than the methylene protons; this is opposite to that expected from Chart 1.

F137 Therefore the n.m.r. absorption corresponding to H^a will appear as a *singlet/triplet*.

A136 insignificant. Coupling is normally only significant in saturated systems when protons are separated by three or fewer bonds.

F138 However, the two identical protons H^b come under the influence of the three identical protons H^c. In the absence of H^c (e.g. in the trichloro derivative $CH_3COOCH_2CCl_3$) the signal for H^b would be observed as a *singlet/triplet/doublet/quartet*.

A137 singlet.

F139 In ethyl acetate itself, the resonance position for H^b will depend on the orientations of the three protons H^c with respect to the applied field. These orientations can be written in eight ways as shown below:

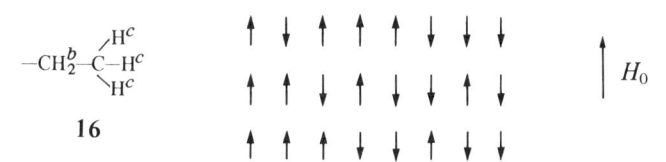

$$-CH_2^b-C\begin{matrix}H^c\\-H^c\\H^c\end{matrix}$$

16

↑↓↑↑↑↓↓↓
↑↑↓↑↓↓↑↓
↑↑↑↓↓↑↓↓   ↑ H_0

These eight combinations can be reduced to four groups which differ from one another with respect to the applied field. They are grouped accordingly below. Fill in the missing columns.

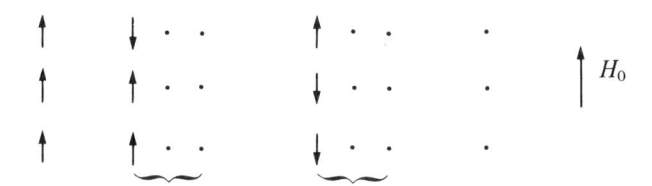

↑   ↓ · ·   ↑ · ·   ·
↑   ↑ · ·   ↓ · ·   ·   ↑ H_0
↑   ↑ · ·   ↓ · ·   ·

A138 singlet. The methyl protons in the trichloro derivative are too far away for coupling to be observed.

F140 Consequently, the signal for H^b will appear as a quartet with area ratios 1:3:3:1 (Fig. 3.10). Now consider the signal due to the methyl protons H^c. At any one instant, their resonance position will be dependent on the orientation of the two protons H^b with respect to the applied field. Write down all

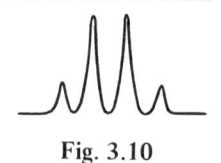

Fig. 3.10

A139

↑   ↓ ↑ ↑   ↑ ↓ ↓   ↓   ↑ H_0
↑   ↑ ↓ ↑   ↓ ↑ ↓   ↓
↑   ↑ ↑ ↓   ↓ ↓ ↑   ↓

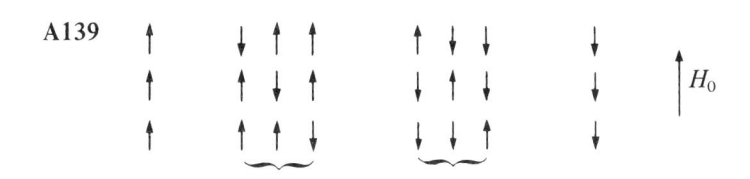

F141 Thus the signal due to H^c will appear as:

(a) a doublet with area ratio 1:1;
(b) a triplet with area ratio 1:2:1;
(c) a triplet with area ratio 1:3:1.

A140

↑ ↓ ↑ ↓   ↑ H_0
↑ ↑ ↓ ↓

F142 Study your conclusions concerning ethyl acetate and decide, using your knowledge of chemical shifts (and Chart 1 if necessary), which of the spectra in Fig. 3.11 corresponds to that of ethyl acetate:

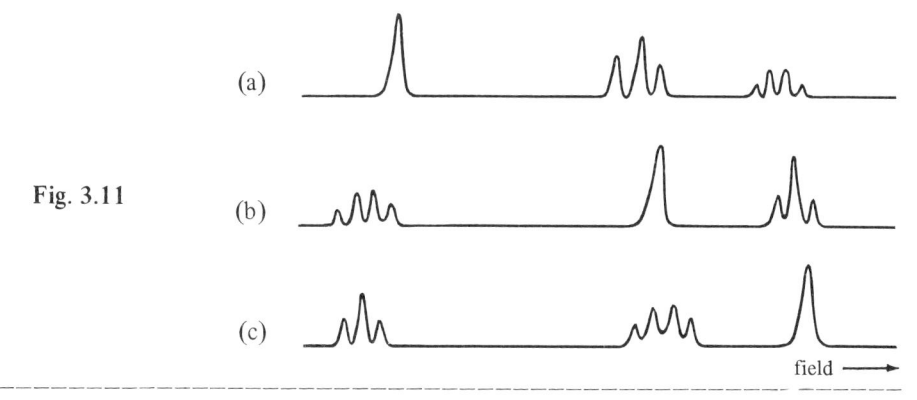

Fig. 3.11

(a)
(b)
(c)

field ⟶

A141 (b) a triplet of area ratio 1:2:1.

F143 You have seen that spin–spin coupling to a single proton results in a doublet; to two identical protons, a triplet; to three identical protons, a quartet. In fact a general rule for the splitting of a signal by spin–spin coupling can be applied as follows. *When a proton interacts with a group of n equivalent protons it will give a signal with n + 1 components.*

The n.m.r. spectrum of 2-bromopropane (**17**) consists of two signals. Write down the number of components into which each is split.

$$\begin{matrix} CH_3 \\ CH_3 \end{matrix} CHBr$$

17

A142 (b). You have deduced in the last few frames the shapes of the signals of the three types of protons. Chart 1 shows that the field required to bring each into resonance increases in the order $-OCH_2-$, CH_3CO, $-CCH_3$.

F144 List the number of components in each of the signals arising from H^a, H^b, H^c, H^d, in molecule 18.

$$\begin{array}{c} ClCH_2^b \\ \diagdown \\ CH^a-C \\ \diagup \quad \diagdown \\ ClCH_2^b \quad \quad \end{array} \begin{array}{c} O \\ \diagup \diagup \\ CH_2^c-CH_3^d \end{array}$$

18

A143 CH_3 protons – 2 components (doublet)
CHBr proton – 7 components (septet)

A144 $H^a - 5, H^b - 2, H^c - 4, H^d - 3$.

The Branching Method

F145 A useful graphical approach to the determination of spin–spin coupling patterns which does not involve working through the number of different orientations of a set of spins can be made by considering the effect of each coupled proton consecutively in the following way.

The aldehydic proton H^x of chloroacetaldehyde (**19**) is coupled to the protons of the methylene group. H^a and H^b are magnetically equivalent and the coupling constant with H^x is the same for each. Nevertheless let us consider them one by one. H^a will split H^x into two components as shown in the figure. H^b will split these components each into two in the same way. As $J_{ax} = J_{bx}$ the central lines coincide indicating that the H^a will appear as a triplet.

Check that you can use the technique to predict the quartet signal of the aldehydic proton of acetaldehyde.

F146 This method is more informative than the $(n + 1)$ rule, in that the relative intensities of the components are deduced in addition to the multiplicity of the splitting pattern. You recall that the ratio of a doublet is 1:1. If in branching diagrams the intensities where the lines coincide are added, the combined ratios can be deduced. Thus the relative intensities of the peaks of a triplet and quartet are derived as shown.

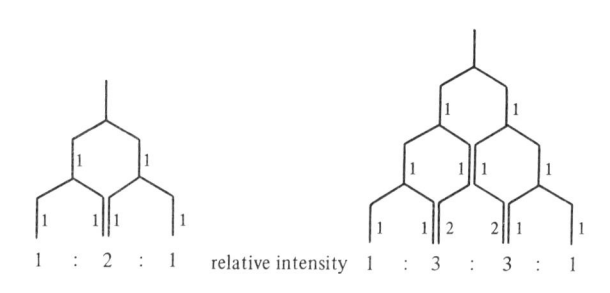

1 : 2 : 1   relative intensity   1 : 3 : 3 : 1

Derive the relative intensities of the peaks in the septet of the methine proton of the isopropyl group.

A145

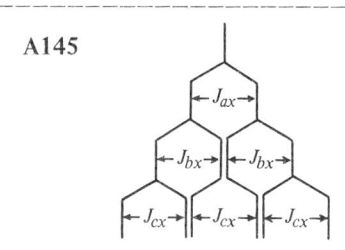

(branching diagram showing J_{ax} at top, J_{bx} in middle row, J_{cx} in bottom row)

F147 Another valuable feature of the branching method is the way in which it can cope with slightly more complicated situations, e.g. where a proton is coupled to other protons which are not equivalent.

Take the hypothetical case of the H^b proton in the system =CH^a–CH^b–CH^c= in which $J_{ab} \gg J_{bc}$. This time none of the lines coincide and the resulting pattern is:

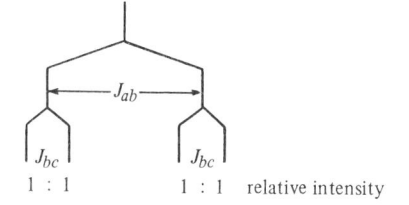

(branching diagram showing J_{ab} splitting then J_{bc} splittings, 1:1 and 1:1 relative intensity)

i.e. a doublet of doublets.

How would the H^b signal in the above system appear in a spectrum if $J_{ab} = 2J_{bc}$?

A146

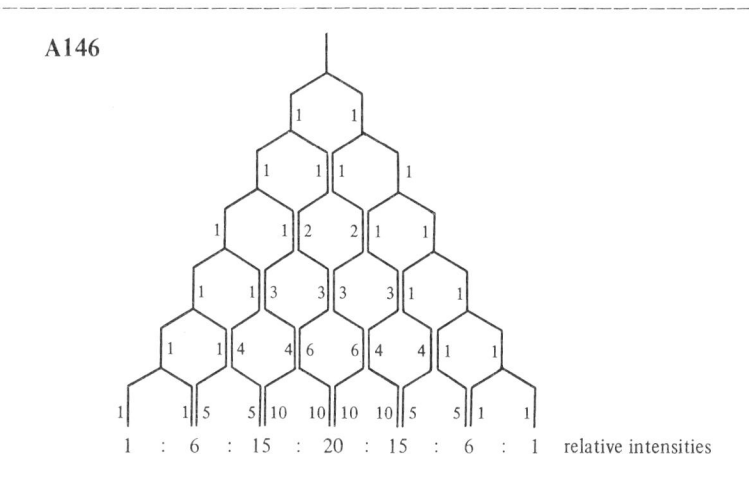

(branching diagram producing relative intensities)

1 : 6 : 15 : 20 : 15 : 6 : 1   relative intensities

F148 The occurrence of an evenly spaced quartet formed in this way is relatively rare. Most quartets observed in n.m.r. spectra arise through the coupling of a proton with:

(a) four equivalent protons;
(b) three equivalent protons;
(c) two equivalent protons.

A147

1 : 1 : 1 : 1 relative intensities

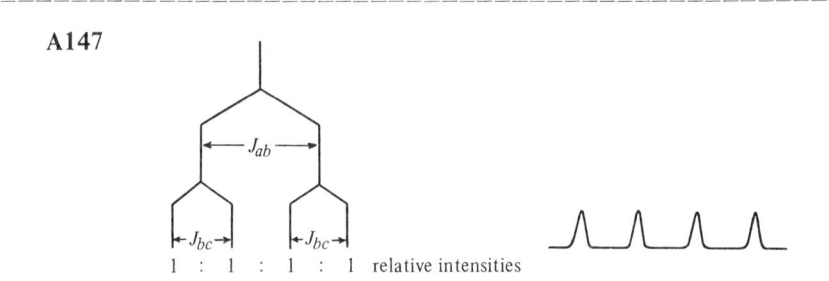

F149 In what way does the quartet arising from coupling with three identical protons differ from the above case in which coupling to only two protons was involved?

A148 (b). The number of components equals the number of equivalent protons plus one (the $n + 1$ rule).

F150 Now consider the case of the signal of the H^b proton in the system shown in **20** in which the coupling between H^a and H^b (J_{ab}) is much smaller than J_{bc} and J_{bd}. Furthermore let $J_{bc} = J_{bd}$ (a common situation) i.e. $J_{bc} = J_{bd} \gg J_{ab}$. Application of the branching technique to the H^b proton leads to the following diagram.

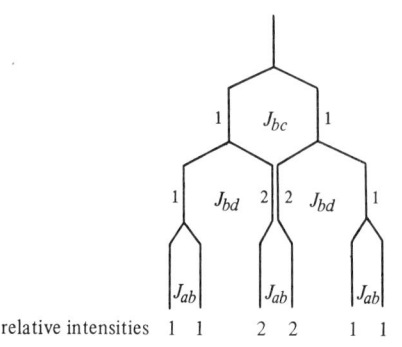

$$\begin{array}{c} H^a \; H^b \; H^c \\ \mid \; \mid \; \diagup \\ -C-C-C-H^d \\ \mid \; \mid \; \diagdown \end{array}$$

20

i.e. a triplet of doublets.

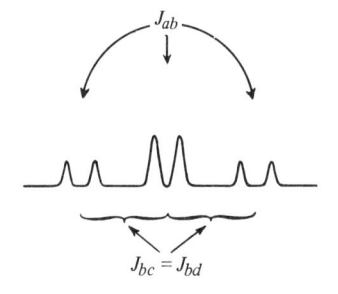

$J_{bc} = J_{bd}$

relative intensities 1 1   2 2   1 1

Use the same approach to work out the shape of the signal when $J_{bc} = J_{bd} \ll J_{ab}$.

A149 The relative intensities of the components of the quartet arising from coupling of a proton with three identical protons is 1:3:3:1, whereas in the $J_{ab} = 2J_{bc}$ case the relative intensities are 1:1:1:1. Thus not only the number of components but also the relative intensities must be taken into account when interpreting an n.m.r. multiplet signal.

F151 The shape of a signal can become very complicated if the restrictions imposed in frame 150 are lifted. For instance, notice how the H^b signal of the system in frame 150 alters when, say, J_{ab} = 10 Hz; J_{bc} = 8 Hz; J_{bd} = 7 Hz. Here all three coupling constants are different, but no one is very different from either of the other two.

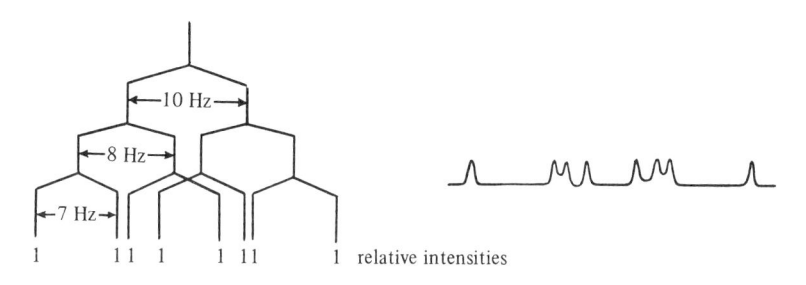

```
        ←―10 Hz―→
     ←―8 Hz―→
  ←7 Hz→
  1   1 1  1    1  1 1   1   relative intensities
```

In general, rapid analysis of a signal will only be easy in simple situations like those described in previous frames. Which of the compounds **21, 22** and **23** is likely to give a very complex signal for the proton indicated?

$$CH_3-\overset{O}{\underset{\|}{C}}-CH_2Cl \qquad \qquad \underset{Cl}{\overset{Cl}{>}}C\underset{C}{\overset{H}{<}}\underset{CH_3}{\overset{H}{<}} \qquad \qquad \underset{CH_3}{\overset{CH_3}{>}}CH-\overset{O}{\underset{\|}{C}}-OCH_3$$

21   **22**   **23**

A150

i.e. a doublet of triplets.

```
         ←―Jab―→
     Jbc
   Jbd
  1:2:1       1:2:1   relative intensities
```

$J_{bc} = J_{bd}$

←―J_{ab}―→

F152 It is not uncommon to observe deceptively simple spectra which arise because of fortuitously equal coupling constants with non-equivalent protons. This can be illustrated for the case considered in frame 150.

$$\underset{R}{\overset{R}{>}}H^a-C-CH^b-C\underset{H^d}{\overset{H^c}{<}}$$
$$RR$$

24

Let $J_{ab} = J_{bc} = J_{bd}$ and apply the branching technique to H^b in the usual way to obtain the relative intensities of the final multiplet.

A151 22. In **22** the proton indicated is coupled to three different types of proton: methyl, methylene on the same side of the ring and methylene on the opposite side of the ring. Each of these is likely to give rise to different coupling constants with the methine proton. The signal in **21** would be a singlet, in **23** a septet.

F153 One can often recognize the presence of fortuitously equal coupling constants by examining the rest of the spectrum.

In which of the spectra in Fig. 3.12 is the quartet (1:3:3:1) produced from fortuitously equal coupling constants involving magnetically non-equivalent protons?

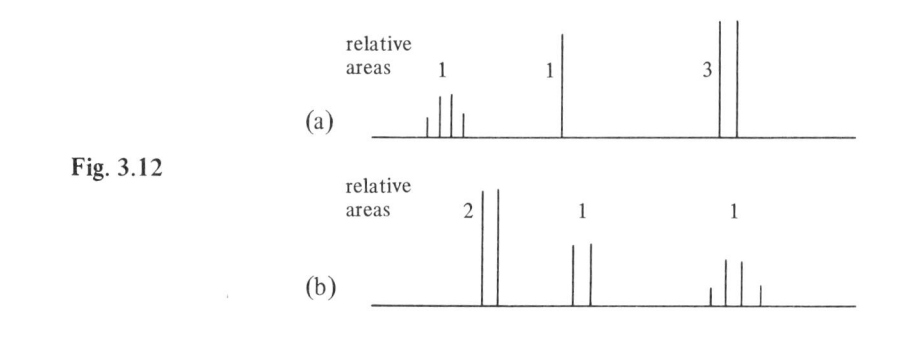

Fig. 3.12

(a) relative areas 1, 1, 3

(b) relative areas 2, 1, 1

A152

J_{ab}, J_{bc}, J_{bd}

1 3 3 1

i.e. a 1:3:3:1 quartet.

F154 State whether the following statements are true or false.
(a) The absorption signal due to a proton coupled to four identical protons is a quintet.
(b) All methyl groups appear as triplets in n.m.r. spectra.
(c) A proton coupled to one other proton always gives rise to a doublet except when the protons have identical chemical shifts.
(d) A proton coupled to three other protons always gives rise to a quartet.

A153 (b). Quartets normally occur in spectra as a result of coupling with three identical protons. Spectrum (b) does not possess a signal due to three identical protons so the quartet present must be due to fortuitously equal coupling constants.

A154 (a) True. (b) False. The multiplet structure of a signal is governed not by the number of protons in the group, but by the number of protons with which the group is coupled. (c) True. (d) False. Coupling to three protons produces a quartet only when all three coupling constants are the same.

NH and OH Protons

F155 We noted in Part 2 that protons on oxygen and nitrogen differ from protons on carbon in that they are usually involved in hydrogen bonding. This results in chemical shift values which vary according to the condition in the sample tube. This part was introduced by reference to the spectrum of ethanol shown below and in frame 125 (Fig. 3.13). Examine it and state whether there is any evidence of spin–spin coupling between the hydroxyl and methylene protons.

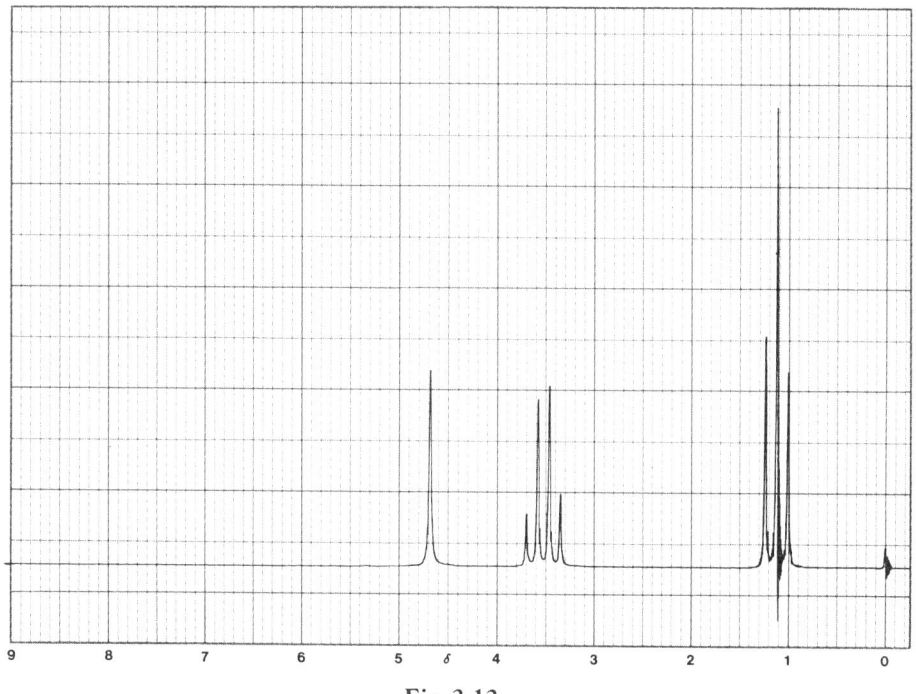

Fig. 3.13

F156 The absence of evidence of spin–spin coupling is a feature common to signals of hydrogen atoms attached to oxygen and nitrogen atoms, when solutions of samples are made up in the usual way. Unless special precautions are taken, protons attached to a heteroatom exchange rapidly and there is insufficient time for a particular OH or NH proton to be 'seen' by neighbouring protons and vice versa. Thus, a singlet is recorded for the exchangeable proton and shapes of signals due to neighbouring protons are unaffected. The appearance of a singlet for the ethanol hydroxyl proton, and a quartet for the methylene protons is thus explained. Write down the multiplicities expected for each signal of compounds **25** and **26** assuming sample conditions to be similar to those for the ethanol spectrum.

$$H^a\!\!-\!\!\underset{H^a}{\overset{H^a}{C}}\!\!-\!\!\underset{H^b}{\overset{Cl}{C}}\!\!-\!\!Cl \qquad H^c\!\!-\!\!\underset{H^c}{\overset{H^c}{C}}\!\!-\!\!O\!\!-\!\!H^d$$

         25   **26**

A155 No. The OH absorption is a sharp singlet and the CH_2 absorption is a 1:3:3:1 quartet due to coupling with the CH_3 group only.

F157 If the rate of exchange of an OH proton is made very slow, e.g. by removal of trace quantities of acid or base impurities, then the normal coupling effects can be observed. The spectrum in Fig. 3.14 is that of a very pure sample of ethanol.

Notice the effect of slow exchange, by comparing this spectrum with the spectrum in Fig. 3.13.

Does the spectrum of $C_6H_5C{\equiv}CCH_2CH_2OH$ shown in Fig. 3.15 indicate that rapid exchange of OH protons is taking place?

A156 H^a 1:1 doublet
H^b 1:3:3:1 quartet
H^c singlet ⎫ The OH proton is exchanging too fast to be 'seen' by
H^d singlet ⎭ the methyl protons.

F158 Exchange can also occur in samples of compounds with protons attached to nitrogen. At rapid exchange rates a proton is not located on the nitrogen atom long enough to 'see' spin states of neighbouring protons, and coupling is not observed. Given that under normal conditions most aliphatic amines undergo rapid NH exchange, which of spectra (a), (b) and (c) in Fig. 3.16 corresponds to methylamine?

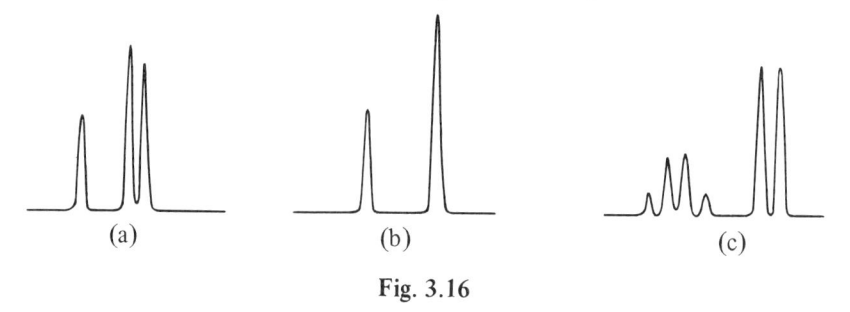

Fig. 3.16

A157 Yes, the two triplets are those of the two methylene groups, and the aromatic protons absorb between 7 and 8 p.p.m. The remaining signal at 4.9 p.p.m., due to OH, shows no signs of spin–spin coupling.

F159 At slow exchange rates, the NH proton remains on the nitrogen atom long enough to be 'seen' by adjacent CH protons. Coupling therefore takes place under such conditions. For example, the signal due to protons of the methyl group in a compound CH_3-NH-R would appear as: (a) a triplet; (b) a doublet; (c) a singlet.

A158 (b). Rapid exchange prevents the NH protons from being 'seen' by the methyl protons and vice versa and so coupling is not observed.

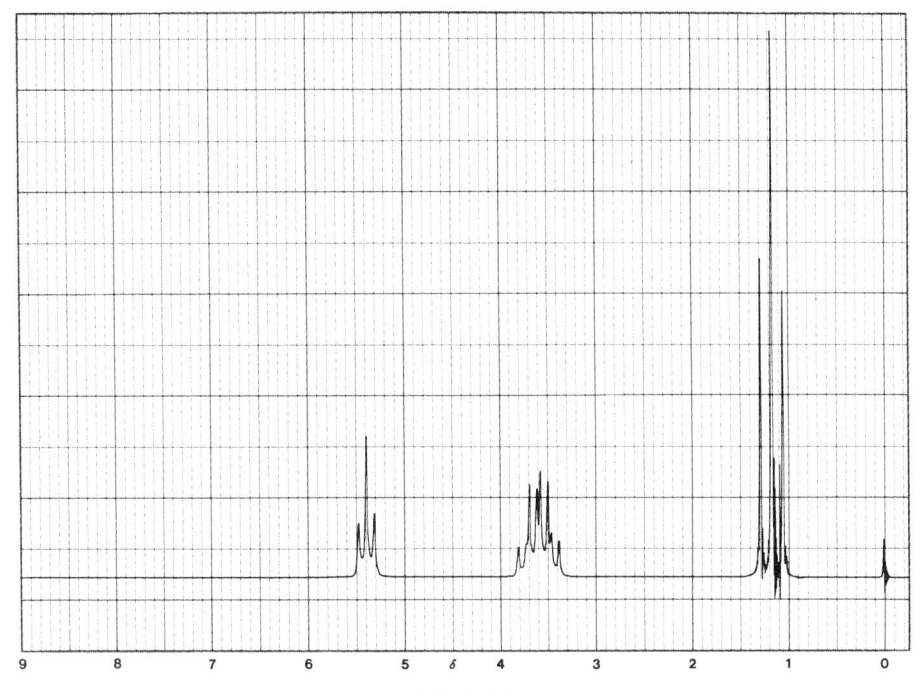

Fig. 3.14

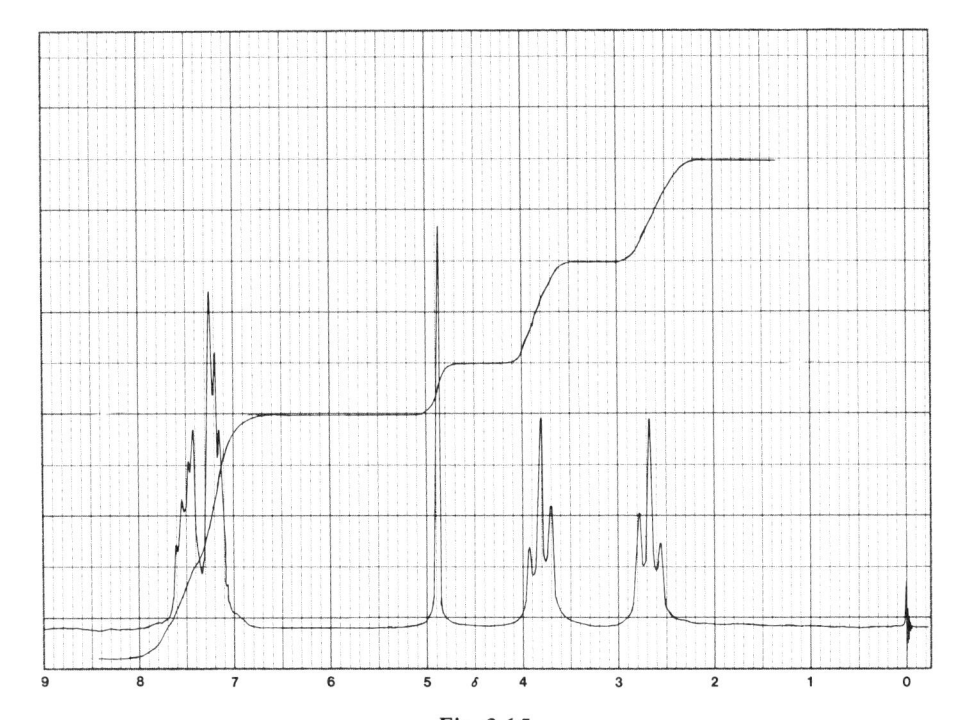

Fig. 3.15

F160 Even when exchange is slow, the signal due to the NH proton rarely appears as a multiplet. Its signal is affected by the magnetic and electric properties of the ^{14}N nucleus in such a way that it usually appears as a broad absorption. Which of the signals A, B, C or D in Fig. 3.17 correspond to protons x and y in the spectrum of the compound $CH_3^x NH^y COOCH_2 CH_3$?

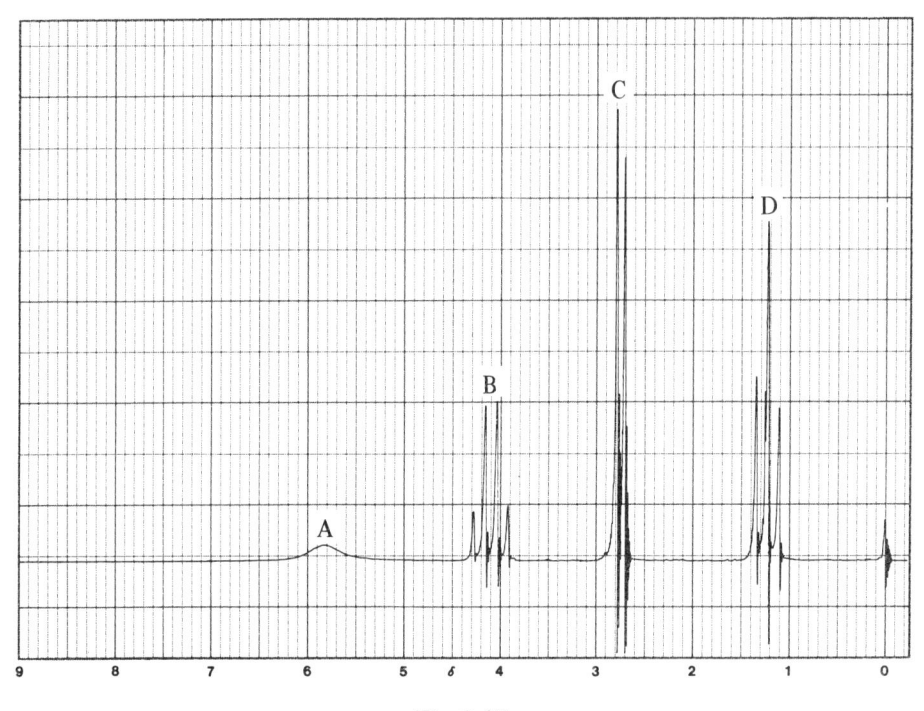

Fig. 3.17

A159 (b) a doublet.

F161 We therefore have the following situation:

	Appearance of NH signal	*Appearance of neighbouring CH signal*
Fast exchange	Sharp singlet	No coupling to NH observed
Slow exchange	Broad band	Coupling to NH observed

One also encounters a third situation intermediate between these two in which exchange is slow enough to show a broad signal for the NH proton but too fast for coupling to neighbouring CH protons to be observed.

Which of the partial spectra in Fig. 3.18 could correspond to that of compounds of the type $RCOCH_2 NHR'$ under circumstances of: (a) fast exchange; (b) slow exchange; (c) intermediate exchange?

(i) (ii) (iii) (iv)

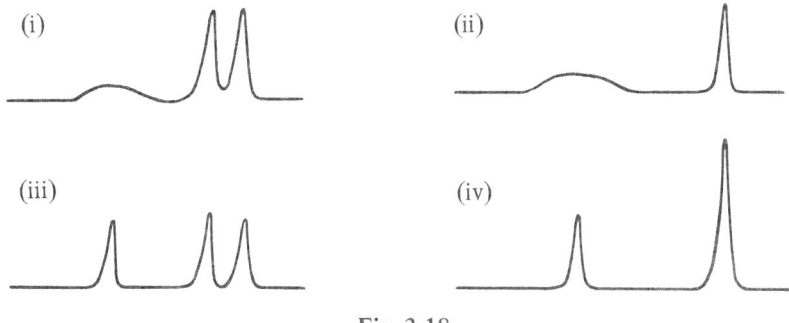

Fig. 3.18

A160 $x = C$; $y = A$. The broad band A is characteristic of slowly exchanging NH. Coupling with adjacent protons takes place under such conditions and the signal of the neighbouring methyl protons appears as a doublet.

A161 (a) = (iv); (b) = (i); (c) = (ii); (iii) shows fast exchange and coupling to neighbouring CH_2 at the same time!

Examples
F162 Problem 1
The spectrum shown in Fig. 3.19 is that of a compound $C_4H_6Cl_2O_2$ which, from infrared evidence, appears to be an ester. First measure the integration curve and deduce the number of vertical units per proton.

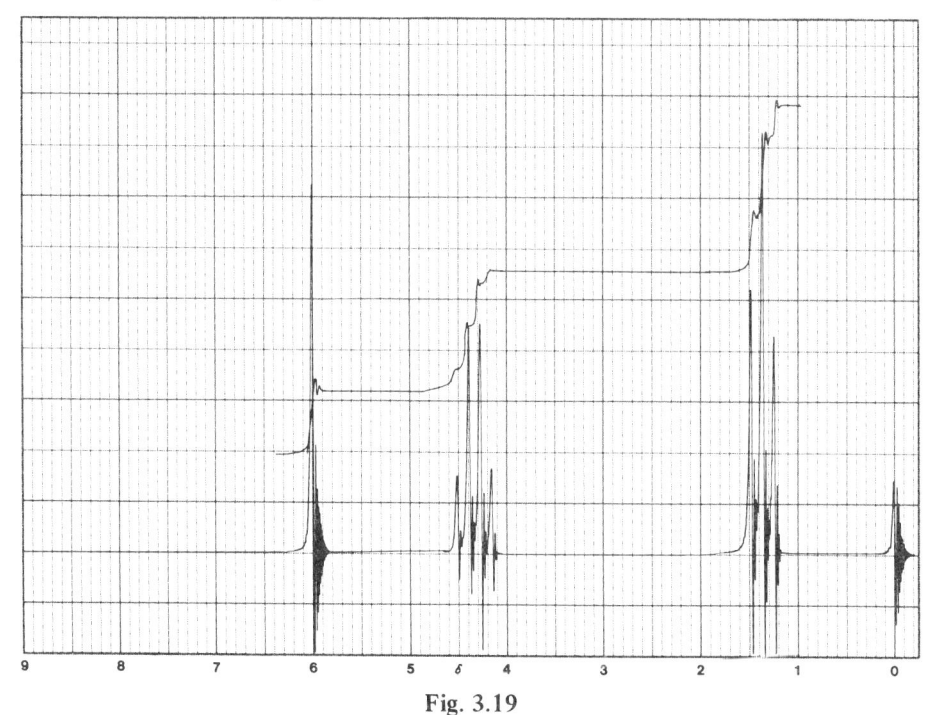

Fig. 3.19

F163 There are three main signals. How many protons correspond to each signal?

> **A162** 1.15. The total number of squares measured vertically between the top and bottom horizontal parts of the integration curve is 6.9. Therefore the number of squares per proton is 6.9/6.

F164 Consider the two multiplets. Which of the following statements is true?
(a) They must represent two sets of protons which are coupled to one another.
(b) They do not represent two sets of protons which are coupled to one another.
(c) They need not represent two sets of protons which are coupled to one another.

> **A163** Singlet, 1 (1.3/1.15); quartet, 2 (2.3/1.15); triplet, 3 (3.25/1.15).

F165 Which kind of group is giving rise to this type of coupling pattern?

> **A164** (a). This is because there are only two multiplets in the spectrum. Notice the slight distortions of the shapes of these signals; they both 'lean' towards one another. This is useful in searching for coupled partners when three or more sets of multiplets are present. Of course, coupled partners must also show equal coupling constants (see frame 129).

F166 Now consider the chemical shift of the $-CH_2-$ group and using Chart 1, suggest possible atoms or groups to which it could be attached.

> **A165** $-CH_2CH_3$. The quartet signifies that the two protons responsible are coupled to three identical protons. This quartet/triplet pattern is characteristic of an ethyl group.

F167 One of these can be eliminated. Which one and why?

> **A166** $-O-$; $-$halogen.

F168 So the unknown ester contains an ethoxy group, (C_2H_5O-). This leaves C_2HCl_2O to account for. Write down the only possible structure that fits these data.

> **A167** $-$halogen. This would give ethyl chloride as the unknown compound with C_2HClO left over!

F169 How do you explain the high δ-value for the methine proton?

> **A168** $CHCl_2-CO-O-CH_2-CH_3$.

A169 Chart 1 indicates that —CH—Hal can absorb at up to 5.8 p.p.m. This particular methine proton has two neighbouring chlorine atoms and a carbonyl group. The combined electron withdrawing effects would be expected to lead to a high δ-value.

F170 Problem 2

Infrared spectroscopy indicates that a compound $C_9H_{15}NO_5$ contains carbonyl but no alkenic bonds. Its n.m.r. spectrum is given in Fig. 3.20. Determine how many protons are responsible for each of the five absorptions (A to E).

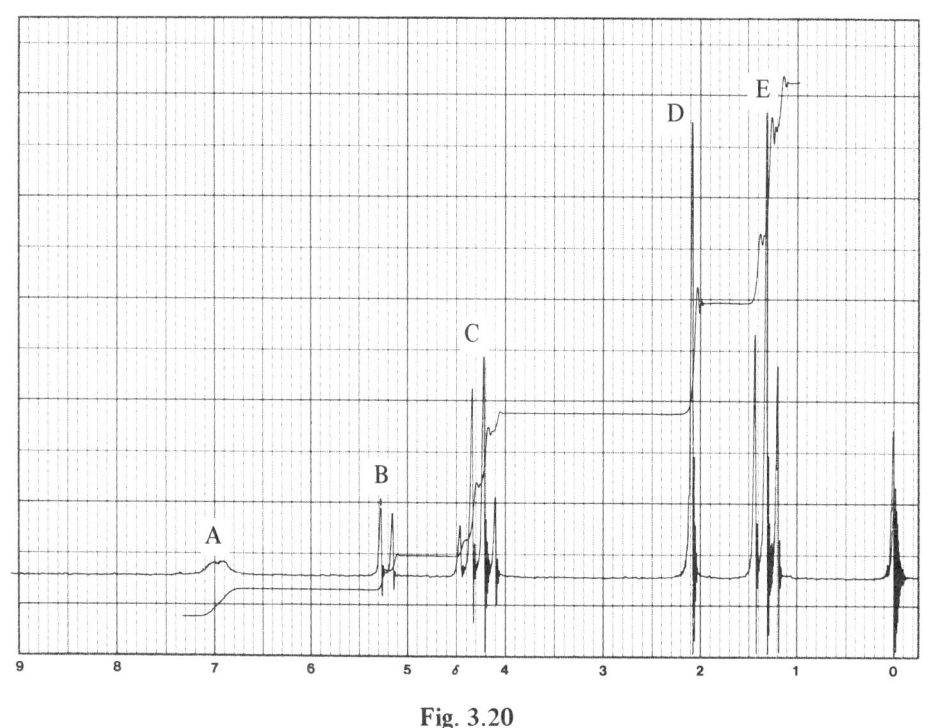

Fig. 3.20

F171 Three of the absorption bands clearly exhibit coupling. Find two which represent coupling partners remembering that slight distortion of multiplets always occurs in which the intensities 'lean towards' one another (see frame 164).

A170 1 : 1 : 4 : 3 : 6
          A   B   C   D   E

F172 Look at the coupling patterns and decide what grouping is responsible for the absorptions C and E, bearing in mind their relative intensities are 4:6.

A171 C and E lean towards each other. B, on the other hand, is a slightly distorted doublet leaning away from both C and E.

F173 Use Chart 1 to determine the environment of the methylene groups responsible for signal C.

A172 Two identical $-CH_2CH_3$ groups.

F174 The sharp singlet at 2.1 p.p.m. (D) represents absorption due to three identical protons and is almost certainly due to a methyl group. Use the chemical shift value to suggest the nature of the group to which it is attached.

A173 $-OCH_2CH_3$. The chemical shift for signal C (centre of the multiplet) is 4.3 p.p.m. The only methylene protons with this value are those attached to halogen or oxygen. The former is eliminated by the molecular formula.

F175 Two bands remain, each due to one hydrogen atom. Coupling between them is indicated because the 5.2 p.p.m. band is a clearly defined doublet with a distortion leaning towards the broad band at 7.0 p.p.m.

Draw a molecular fragment containing four atoms which will explain both the breadth of the 7.0 p.p.m. absorption and the doublet at 5.2 p.p.m.

A174 $-\underset{\underset{O}{\|}}{C}-$. Remember C=C bonds are absent (frame 170).

F176 Formulate a possible structure for the unknown compound bearing in mind the molecular formula ($C_9H_{15}NO_5$) and that so far we have accounted for two equivalent $-OC_2H_5$ groups, a CH_3CO group and a $-NHCH-$ group.

A175 $\underset{}{-N-C-}$ with H H above and a bond below on C. The broad band is characteristic of NH protons undergoing slow or intermediate exchange. Slow exchange would lead to coupling with an adjacent methine proton being observed as a doublet (see frame 161).

A176

$$CH_3-\underset{\underset{O}{\|}}{C}-NH-CH\underset{\diagdown}{\overset{\diagup C-O-CH_2-CH_3}{}}\underset{\underset{O}{\|}}{C}-O-CH_2-CH_3$$

28

F177 Problem 3

The spectrum shown in Fig. 3.21 is that of a compound $C_5H_8O_2$ which from other information appears to be a lactone, that is a cyclic ester of the form shown in 28. You will notice that the spectrum contains a clearly defined doublet (or a pair of singlets) (A) near 1.4 p.p.m., an apparent sextet in the region of 4.7 p.p.m. (C), and a complex absorption pattern in the region 1.6 to 2.8 p.p.m. (B). The interpretation of this complex pattern is likely to be extremely difficult, so initially let us confine our attention to the simple features of the spectrum and use the complex region only as a contribution to the integration curve.

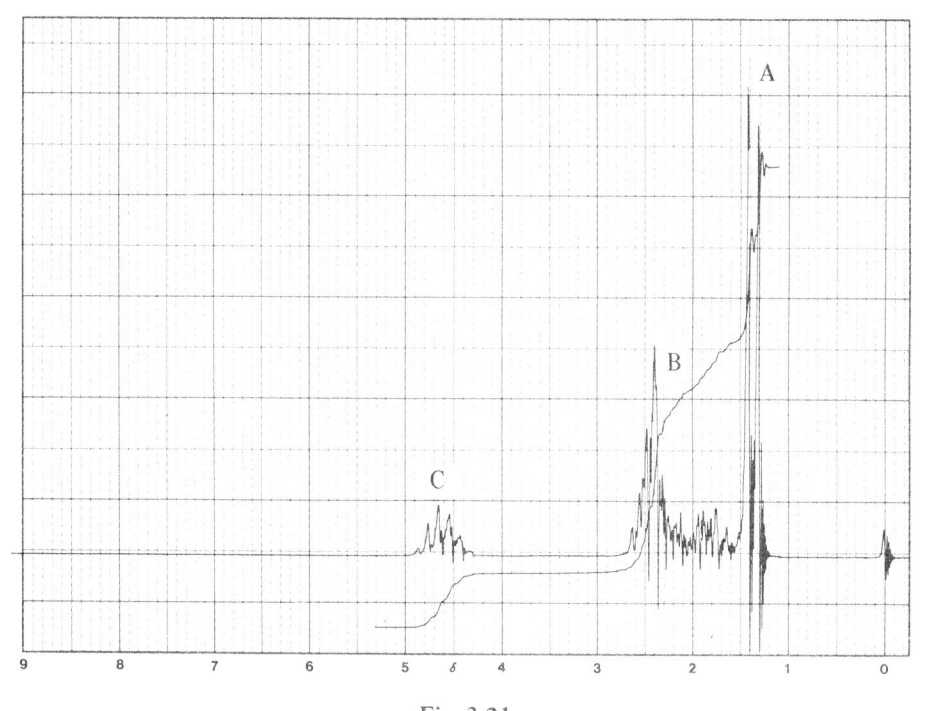

28

Fig. 3.21

Determine the number of protons associated with the regions A, B and C.

F178
Use this information to draw the three possible lactones which fit the molecular formula.

A177
A: three protons; B: four protons; C: one proton.

F179
Each lactone contains a methine proton distinguishable by its δ-value. Examine the spectrum in Fig. 3.21 and the chemical shift chart and deduce which structure is correct.

A178

29: CH₂–CH(CH₃)–O–C(=O)–CH₂–CH₂ (ring)

30: CH₃–CH–CH₂–C(=O)–O–CH₂ (ring)

31: CH₃–CH(O–)–C(=O)–O–CH₂–CH₂ (ring)

F180 A final point worth noting is that the methine signal approximates to a sextet. To account for this, the coupling constant between the CH₃ and the CH protons must be *very different from/very similar to* the coupling constant between the CH proton and the neighbouring –CH₂– group.

A179 Chart 1 shows that the only types of methine protons permissable are –CHO– and R\R/CH–halogen. Halogens are absent from the molecular formula. You have deduced the structure of the compound without completing a detailed analysis of the spectrum. This illustrates how limited information from an n.m.r. spectrum, when coupled with other chemical and physical data, can be invaluable in structure elucidation.

(structure 31: CH₃–CH–O–C(=O)–CH₂–CH₂ ring)

A180 very similar to. Equal coupling to five protons gives a sextet.

Coupling Constants

F181 So far we have described the usefulness of spin–spin coupling of n.m.r. signals in terms of the easily recognizable multiplets observed in simple cases, and the way in which complex spectra can arise through coupling in complicated cases. At the beginning of this part we mentioned that the coupling constant J is a measure of the interaction between two coupled protons, H^a and H^b. Which of the four quantities, a, b, c or d, represents the J-value of the 1:3:3:1 quartet shown?

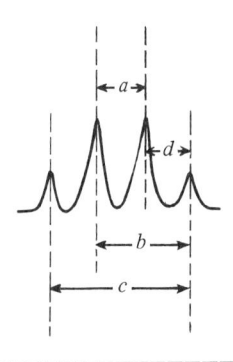

F182 Name the unit in which J is usually measured.

A181 a and d. This can be seen from the branching diagram.

F183 Which of the following statements is correct?

(a) J is dependent on the external applied field and the nature of the links between the two interacting protons.
(b) J is independent of the external applied field, but depends on the nature of the link between the two interacting protons.
(c) J is dependent on the external applied field but independent of the nature of the link between the two interacting protons.

A182 Hertz.

F184 The magnitude of J, then, is dependent only on the nature of the link between the coupled protons. Hence, the value of J revealed in a spectrum can be very useful in structure analysis. The usual range of coupling constants is 0.5 to 20 Hz for proton–proton interaction, and as with chemical shifts, they can be used empirically to recognize structural fragments in a molecule. A typical table is given in Chart 2.*

Pair each of the numerical coupling constants, 3.0, 18.0, 11.2, 7.0 Hz with each of the J values, J_{ed}, J_{df}, J_{ef} and J_{ab} in molecule 32.

32

A183 (b). See frame 129.

F185 Some of the factors affecting the magnitude of the coupling constant can be ascertained from the table of values given. The first is the dependence on the number of bonds between interacting nuclei. Consider a molecular fragment containing **only** C–C single bonds and C–H bonds. In such a fragment:

(a) coupling constants of the order of 12 to 15 Hz arise from coupling between protons separated by 2/3/4/5 bonds.
(b) coupling constants of the order of 2 to 12 Hz arise from coupling between protons separated by 2/3/4/5 bonds.
(c) coupling constants of the order of 1 Hz arise from coupling between protons separated by 2/3/4/5 bonds.

A184 $J_{ed} = 11.2$, $J_{df} = 18.0$, $J_{ef} = 3.0$, $J_{ab} = 7.0$ Hz.

---

*Next to Chart 1 at the back of the book.

F186 Use Chart 2 to deduce which of the following halobenzenes would be expected to exhibit spectra showing coupling constants of about 2.5 Hz.

      33   34   35

A185 (a) 2; (b) 3; (c) 4.

F187 The two previous frames demonstrate that coupling falls off rapidly with distance and is not generally observed when interacting protons are more than 3 σ-bonds apart. State whether coupling constants are generally larger or smaller when multiple bonds are introduced into the molecular fragment by comparing values for the following structures.

    H—C—C—H   and   H₂C=CH₂

        36   37

  H—C—C—C—H   and   H—C=C—C—H

        38   39

A186

[structure 34]

Chart 2 indicates that 2.5 Hz is a typical value for a *meta* coupling constant. Both structures **34** and **35** contain *meta* hydrogens, however **35** contains three equivalent protons which will give rise to a sharp single absorption and *meta* coupling is not observed.

F188 Chart 2 shows that the coupling constant for a *cis* disubstituted alkene differs from the *trans* isomer. We have just shown that the magnitude of J is dependent on the nature and number of bonds between interacting nuclei. For *cis* and *trans* isomeric alkenes, however, both the nature and number of interacting bonds is the same. Consequently, there must be another factor which influences the extent of the coupling. There are two possibilities; which do you consider the more likely?   **40**

(a) The distance through space between the protons.
(b) The geometrical orientation of one CH bond relative to the other.

         cis
         trans

    A187 Generally larger.

    A188 (b). We stated earlier (frame 131) that coupling is apparently relayed through bonds and not through space.

Examples
Problems 4 and 5 involve an analysis of the n.m.r. spectra of the two compounds whose structures are known.

F189 Problem 4
The spectrum of ethyl vinyl ether (41) given in Fig. 3.22 has three sets of absorption bands, a triplet around 1.2 p.p.m. (three protons), a rather complex multiplet in the region of 4 p.p.m. (four protons) and a quartet of almost equal intensity peaks around 6.4 p.p.m. (one proton). Clearly the high field multiplet is due to the methyl group, CH_3^e. Write down the chemical shift of the adjacent methylene protons.

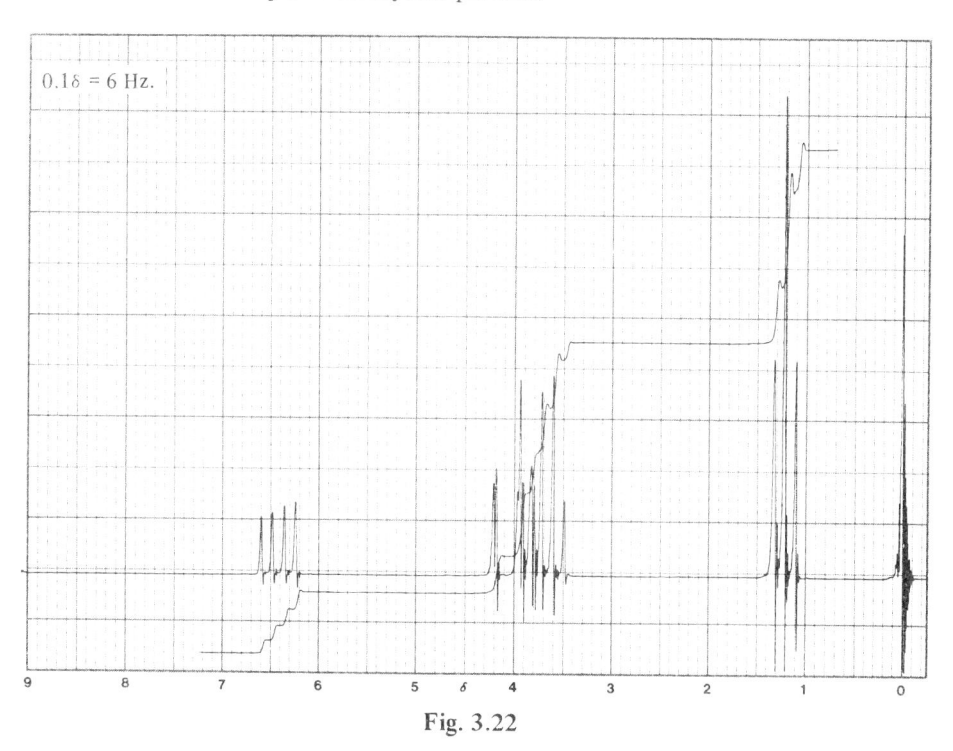

$$CH_3^e CH_2^d O \diagdown C=C \diagup H^c$$
$$H^a \diagup \qquad \diagdown H^b$$
$$41$$

$0.1\delta = 6$ Hz.

Fig. 3.22

F190 An expanded version of the multiplet near 4 p.p.m. is given in Fig. 3.23. Select the components due to the methylene quartet.

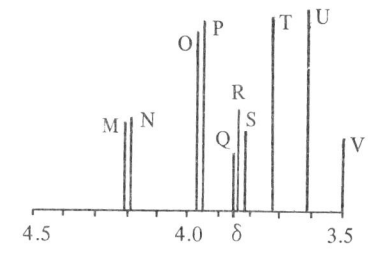

Fig. 3.23

A189 3.7 p.p.m. Three components of the expected quartet are clearly visible on the high field side of the complex multiplet. The chemical shift is taken as the mid-point of the quartet.

F191 You have located all the absorption components of the ethyl group. The rest of the spectrum arises from the alkene protons. In order to focus attention on these absorptions, the methylene quartet has been removed giving Fig. 3.24.

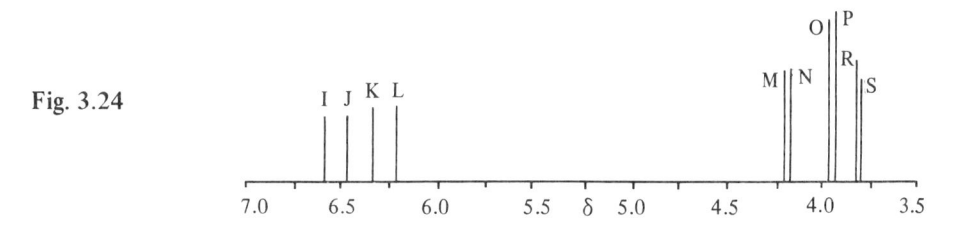

Fig. 3.24

How does this pattern arise? Start by writing down the protons which you would expect to couple with H^a.

A190 Q, T, U, V. Q is the same distance down-field from T as V is upfield from U.

F192 Use the average values for *trans* and for *cis* coupling constants given in the chart to construct a branching diagram for the signal of H^a proton.

A191 H^b and H^c. H^d and H^e are too far away (four and six bonds respectively).

F193 Identify the H^a signal in the spectrum by giving its chemical shift value.

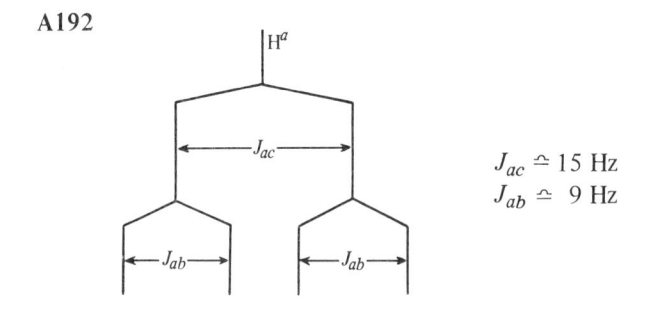

A192

$J_{ac} \doteq 15$ Hz
$J_{ab} \doteq 9$ Hz

F194 The actual shape of the H^a signal is in very good agreement with that predicted by the branching diagram. It would, however, be helpful in making further assignments, to have the precise values of the J_{ab} and J_{ac} coupling constants. Obtain these by measuring the distances between the appropriate peaks of the H^a multiplet on Fig. 3.22.

A193 6.4 p.p.m. The branching diagram predicts four lines of equal intensity for H^a and a chemical shift given by the mid-point of the multiplet.

F195 Predict the multiplet structure of the signal due to H^c (use the J_{ac} value just deduced and an average value for alkene geminal coupling obtained from the chart).

A194 $J_{ab} = 7$ Hz; $J_{ac} = 14.5$ Hz.

F196 Name the labelled components in the simplified spectrum in frame 191 which corresponds to absorption by H^c.

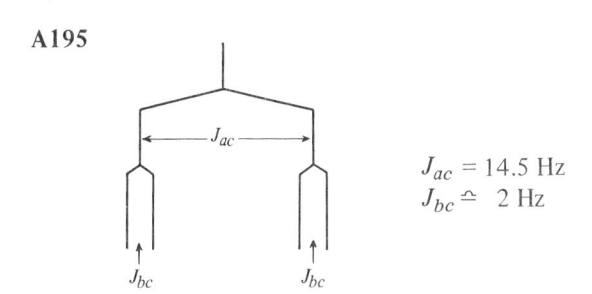

A195

$J_{ac} = 14.5$ Hz
$J_{bc} \triangleq 2$ Hz

F197 The form of the H^b signal should resemble that of the H^c signal (a doublet of doublets) except for a difference in the value of one of the coupling constants. Identify the components in the simplified spectrum in frame 191 which corresponds to absorption by H^b.

A196 M, N, O and P. The branching diagram predicts two doublets separated by 14.5 Hz.

F198 You have now made a full interpretation of the spectrum of ethyl vinyl ether and are in a position to write down values of all the parameters involved. They are:

	H^a	H^b	H^c	H^d	H^e
Chemical shift value (p.p.m.)	6.4	3.9	4.1	3.7	1.2
	J_{ab}	J_{ac}	J_{bc}	J_{dc}	
Coupling constant (Hz)	7	14.5	2	7	

Continue with problem 5.

A197 OP and RS. Note, the low field doublet of the H^b signal and the high field doublet of H^c are coincidentally superimposed upon one another in the OP doublet.

F199 Problem 5

The spectrum of *p*-chlorostyrene (**42**) given in Fig. 3.25 has a similar alkenic pattern to that of ethyl vinyl ether.

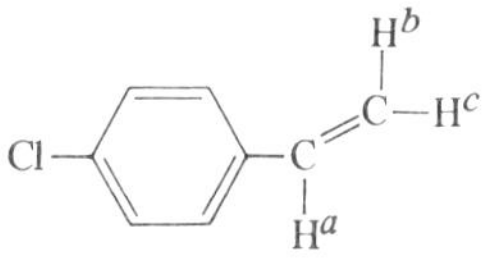

42

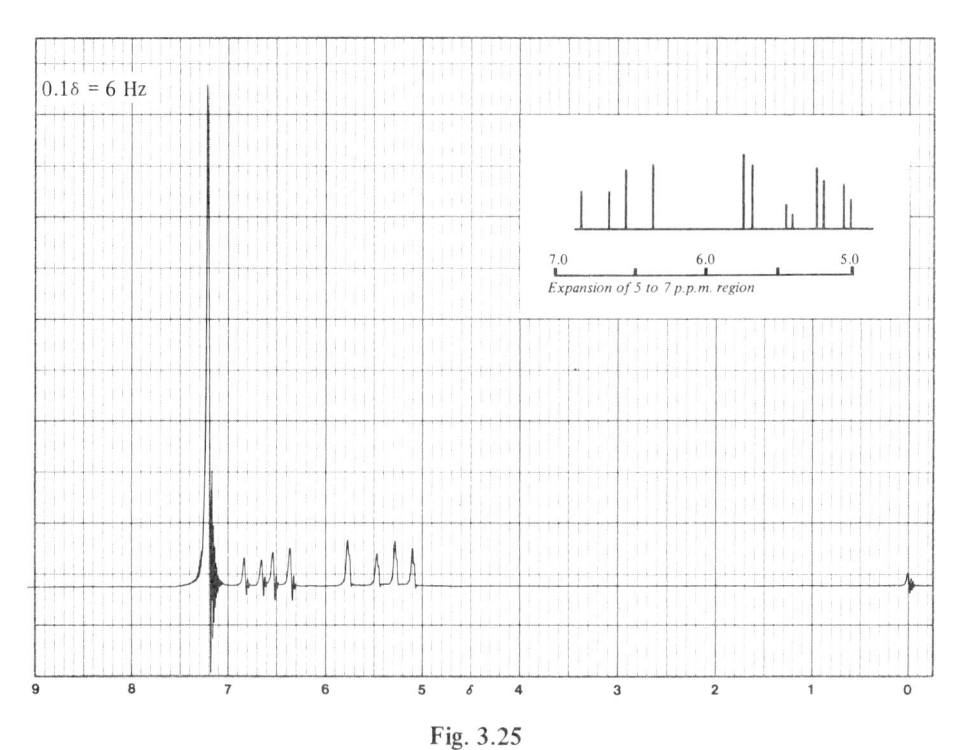

Fig. 3.25

Identify the H^a signal by recording its approximate chemical shift and then use the spectrum to deduce values for J_{ab} and J_{ac} estimating your answers to the nearest whole number of hertz.

F200 Record the approximate chemical shift of H^b and H^c and the value of the J_{bc} coupling constant.

A199 6.6 p.p.m.; J_{ab} = 18 Hz; J_{ac} = 11 Hz. All these values are obtained from the absorption between 6.3 and 7.0 p.p.m. The centre of the multiplet gives the approximate chemical shift of H^a and the two coupling constants are obtained in the following way.

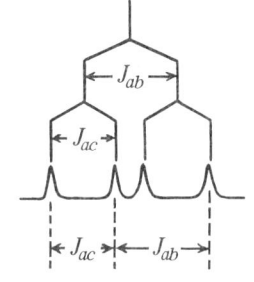

A200 H^b = 5.6 p.p.m.; H^c = 5.2 p.p.m.; J_{bc} = 2 Hz. The signal for H^b should appear as a doublet with a coupling constant of 18 Hz, the components of which are further split by coupling to H^c. This pattern is between 5.4 and 6.0 p.p.m. Similar arguments lead to the δ-value of H^c.

Revision Summary
Spin–spin coupling

In the simple system (**43**) the precise field necessary for H^a to resonate at any one instant will depend on whether H^b is aligned with or against the field. Thus the signal for H^a will appear as a doublet; likewise that for H^b will appear as a doublet. The separation of the components of the doublet is known as the coupling constant and given the symbol J. J-values are measured in hertz.

Extension of this argument to the cases **44** and **45** leads to the signal for H^a being a triplet and quartet respectively. In general, the number of components into which an n.m.r. signal of a proton is 'split' is equal to the number of identical protons to which it is coupled, plus one (the $n + 1$ rule).

The size of the coupling constant is dependent on the link between the interacting protons and falls off rapidly as the number of intervening bonds increases. Normally significant coupling is shown only when there are three or fewer interacting bonds in saturated compounds and four or fewer in unsaturated compounds.

The branching method

A graphical approach to the determination of spin–spin coupling patterns is the use of the branching diagram. This is particularly useful in dealing with cases where the proton is coupled to other protons which are not magnetically equivalent. Thus the proton H^a in the alkene (**46**) would be expected to appear as a doublet of quartets.

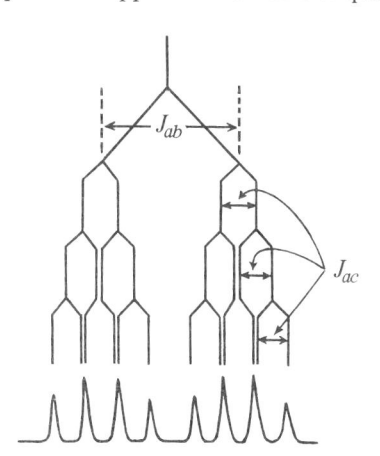

Exchange phenomena

One does not generally observe coupling of hydroxylic protons with protons on neighbouring carbon atoms because hydrogen bonding and rapid exchange does not allow the former to remain on the oxygen atom long enough to be 'seen' by the latter and vice versa. In the absence of exchange normal coupling can be observed. The same applies to NH protons except that one does not observe the NH proton itself as a clearly defined multiplet because of the electrical and magnetic properties of the nitrogen nucleus.

Now try the multiple choice test at the beginning of this part. Write your answers in a vertical column and mark your script as before. Then consult the remarks below.

2–6: Consult your instructor as to whether you should go on or seek extra tuition. If you are using the programme for private study, reread the frames pertinent to those you answered incorrectly and, when you are satisfied you fully understand the material, read the paragraph at the end of this part and start the next part.

More than 6: Revise the material pertinent to those questions you answered incorrectly. If your instructor does not want you to consult the bibliography or attempt any of the questions which follow, read the paragraph at the end of this part and continue with the next part.

Questions

1. Study the spectrum and structure of compound **47** and deduce how the observed spin–spin splitting patterns arise.

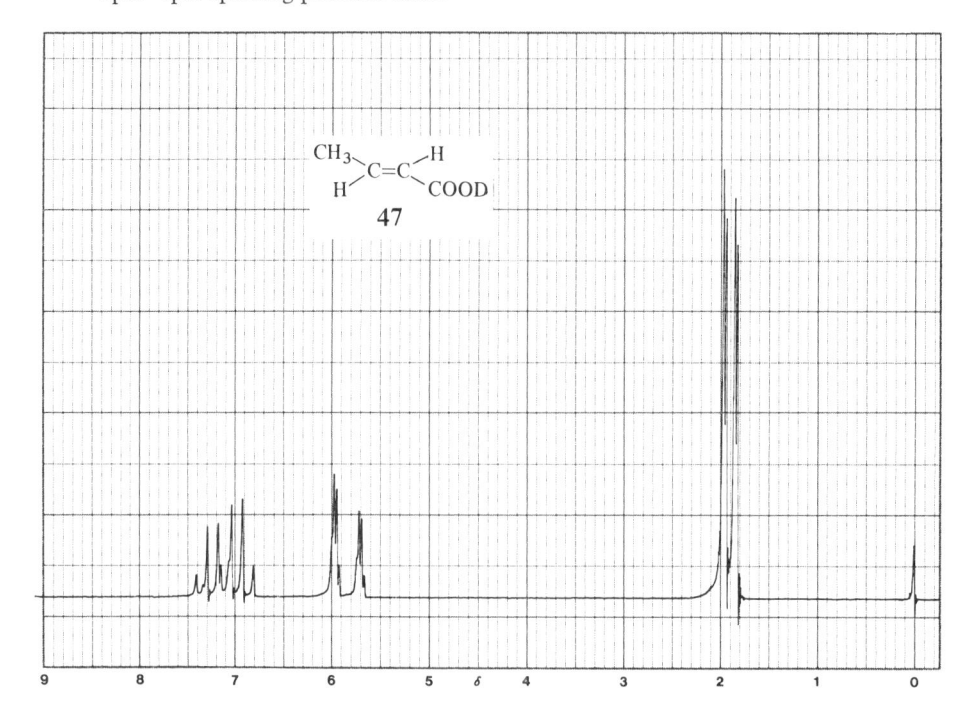

2. Features of the n.m.r. spectra of compounds **48** to **52** are given in (a) to (e). Assign each feature to one of the structures.

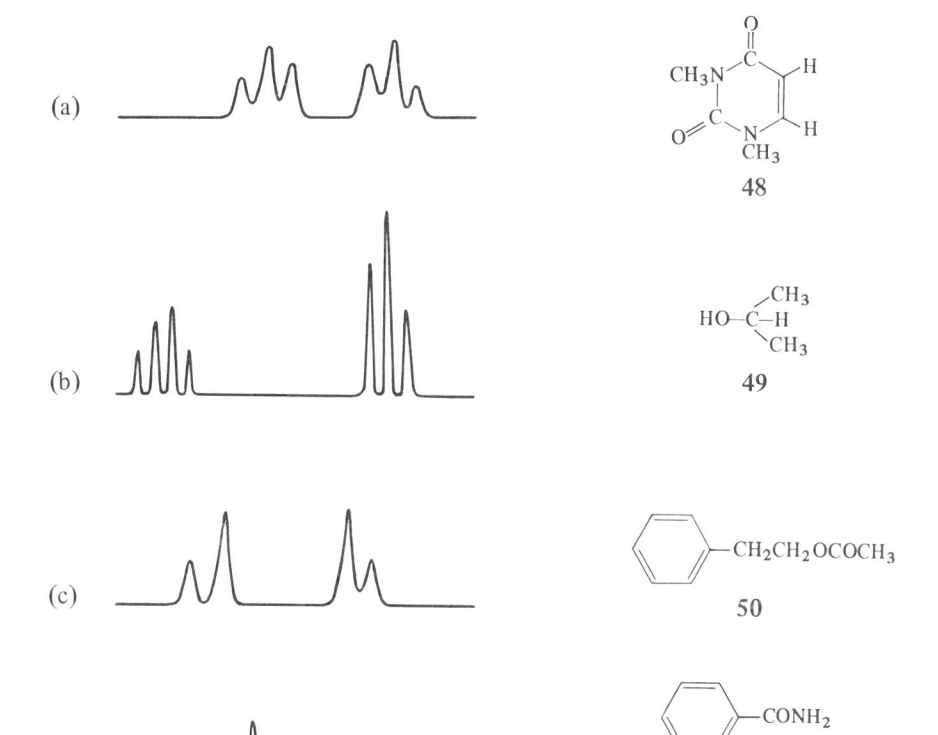

(a)

$$\begin{array}{c} O \\ \parallel \\ CH_3N \diagdown C \diagup H \\ O = C \diagdown N \diagup H \\ | \\ CH_3 \end{array}$$

48

(b)

$$\begin{array}{c} CH_3 \\ HO-C-H \\ CH_3 \end{array}$$

49

(c)

Ph—CH$_2$CH$_2$OCOCH$_3$

50

(d)

Ph—CONH$_2$

51

(e)

Ph—CH$_2$CH$_3$

52

3. Suggest three possible structures for the heterocyclic compound $C_3H_4N_2S$.

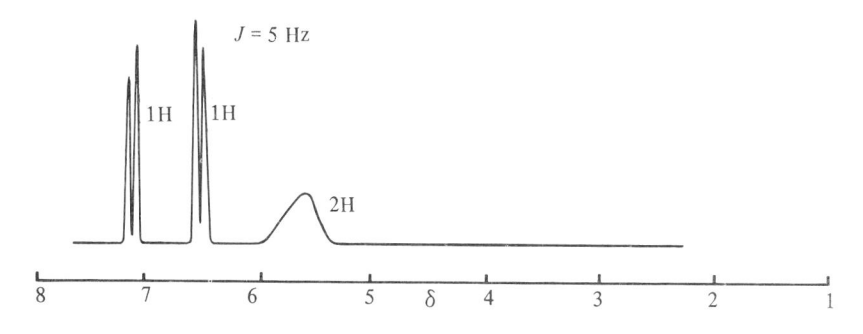

$J = 5$ Hz, 1H, 1H, 2H
(δ axis from 8 to 1)

Further Reading

L. M. Jackman and S. Sternhell, *Applications of Nuclear Magnetic Resonance in Organic Chemistry*, Pergamon Press, Oxford, 1969, Chapters 2.3, 4.1, 4.2, 4.3 and 4.4.

D. W. Mathieson, *Nuclear Magnetic Resonance for Organic Chemists*, Academic Press, London, 1967, Chapters 4, 5, 6, 7 and 8.

J. W. Emsley, J. Feeney and L. H. Sutcliffe, *High Resolution Nuclear Magnetic Resonance Spectroscopy*, Volume 1, Pergamon Press, Oxford, 1965, Chapter 3.

D. Chapman and P. D. Magnus, *Introduction to Practical High Resolution Spectroscopy*, Academic Press, London, 1966.

J. R. Dyer, *Application of Absorption Spectroscopy of Organic Compounds*, Prentice-Hall, New Jersey, 1965, Chapter 4, paragraphs 4, 9 and 12.

R. M. Silverstein and G. C. Bassler, *Spectrometric Identification of Organic Compounds*, John Wiley & Sons, New York, 1967, Chapter 4, paragraphs 4, 5 and 12.

D. H. Williams and I. Fleming, *Spectroscopic Methods in Organic Chemistry*, McGraw-Hill, Maidenhead, 1966, Chapter 4, paragraphs 5, 7, 8 and 9.

R. H. Bible, Jr., *Interpretation of N.m.r. Spectra*, Plenum Press, New York, 1965.

S. F. Dyke, A. J. Floyd, M. Sainsbury and R. Theobald, *Organic Spectroscopy – An Introduction*, Penguin, 1971, Chapter 4, section 4.5.

E. D. Becker, *High Resolution N.m.r.*, Academic Press, New York, 1970, Chapter 5.

F. A. Bovey, *Nuclear Magnetic Resonance Spectroscopy*, Academic Press, New York, 1967, Chapter 5.

W. W. Paudler, *Nuclear Magnetic Resonance*, Allyn and Bacon, Boston, 1971, Chapter 3.

You are now in a position to apply the basic principles of spin–spin coupling in the determination of structural information from n.m.r. spectra. However, coupling can and often does lead to very complex spectra not readily amenable to analysis by direct application of these principles. In Part 4 we consider the circumstances which give rise to complex spectra and the ways in which the chemist obtains information in such cases.

PART 4
COMPLEX (SECOND-ORDER) SPECTRA

Aim
After completing this part of the programme the student will be able to recognize when very complex signals can be expected. Furthermore, he will appreciate the scope of the special methods available (double resonance, use of high field strength, deuterium substitution, computer techniques and chemical shift reagents) for extracting information from complicated spectra. He will be able to use the alphabetical notation to describe spin systems, including cases of chemically equivalent but magnetically non-equivalent protons.

New Terms and Concepts
Chemically equivalent but magnetically non-equivalent protons, the *AMX* notation for spin systems, double resonance, spin decoupling, chemical shift reagents.

Prior Knowledge
Parts 1, 2 and 3; the structure of simple aromatic and aliphatic compounds.

Objectives for Part 4
When you have completed this part, you should be able to:
1. Recognize distortion of coupling patterns related to the comparative magnitudes of δ and J.
2. Interpret spin–spin coupling situations in terms of the standard alphabetical notation.
3. Given possible structural alternatives for an unknown, predict whether or not a particular complex signal in a spectrum will reveal significantly more information if (i) a double irradiation experiment is carried out; (ii) higher field strengths are used; (iii) deuterium labelling or exchange techniques are used; (iv) one had access to computed spectral patterns; (v) a paramagnetic shift reagent is added to the sample.

Introduction
Many n.m.r. spectra show more complex signals than might be expected from applying the principles in Part 3. Part 4 begins by indicating the kind of circumstances which lead to complex spectra and then goes on to summarize some of the methods available for extracting the maximum information in such cases.

Attempt the following multiple choice test, marking it in the usual way, by reference to the correct answer sequence on p. 175 and then compare your marks with the comments set out after the test. Do not seek the correct answers to incorrectly answered questions at this stage.

Multiple Choice Test
1. The absorption pattern in Fig. 4.1 corresponds to two coupled protons in which:

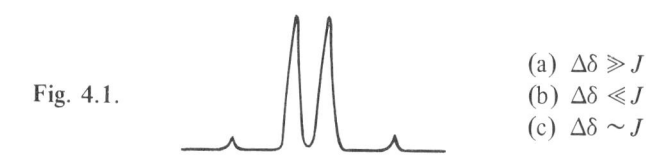

   Fig. 4.1.

   (a) $\Delta\delta \gg J$
   (b) $\Delta\delta \ll J$
   (c) $\Delta\delta \sim J$

2. The protons of an ethoxy group give rise to a characteristic triplet-quartet absorption pattern and the spin system is described by the letters:
   (a) $A_2 B_3$ (b) $A_3 X_4$ (c) $A_2 X_3$

3. Which of the following compounds is most likely to give rise to an $AA' BB'$ spin pattern?

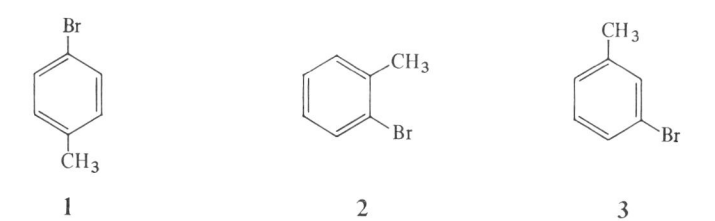

   1: 4-bromotoluene (Br and CH₃ para on benzene)
   2: 2-bromotoluene (CH₃ and Br ortho on benzene)
   3: 3-bromotoluene (CH₃ and Br meta on benzene)

4. For an AB system, mathematical calculation needs to be carried out to determine the value(s) of:
   (a) the chemical shifts, but not the coupling constant;
   (b) the coupling constant, but not the chemical shifts;
   (c) both the chemical shifts and coupling constants.

5. An $A_2 B_2$ type of distorted signal may be simplified to a more recognizable $A_2 X_2$ form by use of:
   (a) an instrument operating at a higher field strength;
   (b) double resonance;
   (c) a computer.

6. The absorption pattern in Fig. 4.2 is one of the signals observed when 1,1,2-trichloropropane ($CH_3^a CH^b ClCH_2^c Cl_2$) is measured whilst being irradiated with an addition frequency corresponding to another signal. Which of the protons (a) a, (b) b, (c) c, has a resonating frequency corresponding to the saturation frequency?

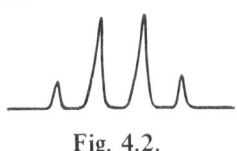

Fig. 4.2.

7. Which of the following coupled spin systems would be readily simplified by the use of double resonance?

   (a) $A_2 B_2$ (b) ABC (c) $A_2 M_2 X$

8. After replacement of the α-protons of a compound of the type $XCH_2^\gamma CH_2^\beta CH_2^\alpha NO_2$ by deuterium to give $XCH_2 CH_2 CD_2 NO_2$ a simplified spectrum can be obtained in which the absorption of:

   (a) the α-protons is absent and that of the β-protons appears as a triplet;
   (b) the α-protons is absent and that of the β-protons appears as a quintet;
   (c) the α-protons appears as a singlet and that of the β-protons as a quartet.

9. The spectrum of an ester (Fig. 4.3) was changed from (I) to (II) by the use of:

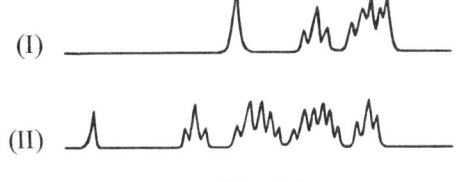

   (I)

   (II)

   (a) $D_2O/NaOD$
   (b) double resonance
   (c) a paramagnetic shift reagent

   Fig. 4.3.

10. In which of the following cases can the overlapping spectral lines be separated by the use of a paramagnetic shift reagent?

    (a) an aliphatic hydrocarbon;
    (b) an aliphatic alcohol;
    (c) an aromatic hydrocarbon.

Comments on marks

Less than 2: You need instruction — continue with Part 4.

3–7: You have some knowledge of this topic but it would be in your interest to secure a more thorough understanding and you are advised to work through Part 4 — it shouldn't take you long!

More than 7: Unless you obtained full marks, consult the sections of Part 4 relevant to your incorrect answers; then proceed directly to Part 5.

PART 4: COMPLEX (SECOND-ORDER) SPECTRA

Complex (Second-Order) Spectra

F201 Until now we have been concerned mainly with coupling patterns where analysis is relatively straightforward; for example the case of two coupled protons has been stated to give a pattern of the type shown in Fig. 4.4.

Fig. 4.4.

Complications arise, however, when the chemical shift difference between the two nuclei ($\delta_A - \delta_B = \nu_{AB}$) is not very much larger than the coupling constant J. The first sign of departure from predicted patterns such as the one given above (first-order splitting) is the distortion of signals which we have encountered already. Which of the two spectra shown in Fig. 4.5 are observed in such cases?

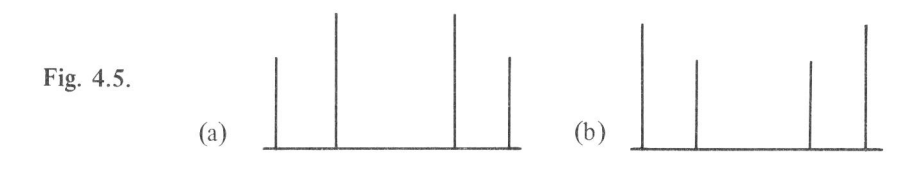

Fig. 4.5.

(a)   (b)

F202 The distortion of the spectra becomes increasingly obvious as $\delta_A - \delta_B$ ($= \nu_{AB}$, the chemical shift difference expressed in hertz) becomes progressively smaller relative to the coupling constant J_{AB} (see Fig. 4.6).

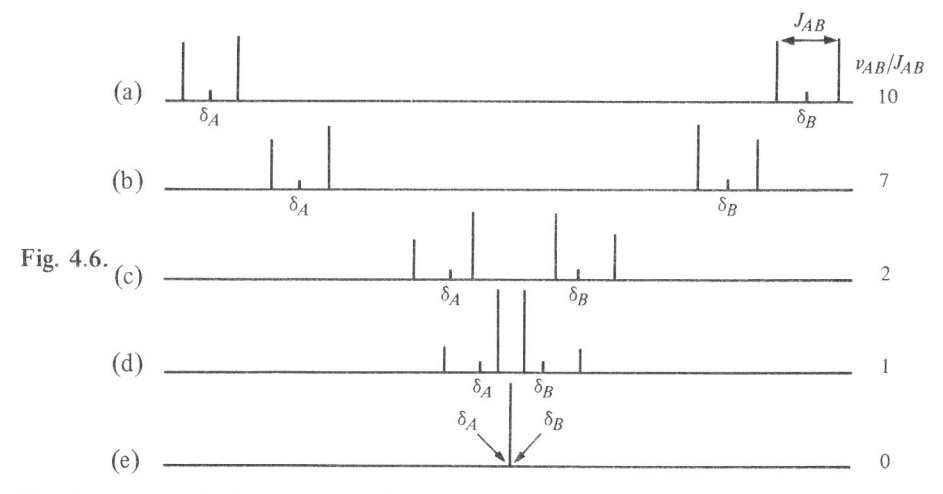

Fig. 4.6.

(a) ν_{AB}/J_{AB} = 10
(b) 7
(c) 2
(d) 1
(e) 0

The distortion, which increases as the ratio ν_{AB}/J_{AB} decreases, also leads to the chemical shift being located closer to the more intense component rather than the mid-point of the doublet. Which ν_{AB}/J_{AB} values correspond to cases where the chemical shift can be determined by visual inspection?

A201 (a). Remember that coupled multiplets lean towards each other as in (a) (see answer 164).

F203 In which of the following cases do the outer components of the doublets disappear?

(a) $(\delta_A - \delta_B)$ is very small; ν_{AB} is very small.
(b) $(\delta_A - \delta_B) = 0$; $\nu_{AB} = 0$.
(c) $(\delta_A - \delta_B) \gg J$; ν_{AB}/J_{AB} is large (say > 10).

A202 >10 and 0. In (a) $\nu_{AB}/J_{AB} > 10$, the distortion is very small and the chemical shift lies virtually at the mid-point of each doublet. (e) corresponds to the case when the chemical shifts are the same; note that coupling still takes place but cannot be observed directly. Of the intermediate cases, (b) shows only a small distortion and an estimation can be made to a very good approximation. In (c) and (d), however, visual inspection is inadequate for the determination of accurate chemical shifts and a full mathematical analysis would be necessary.

F204 Look again at the spectra in frame 202. In which cases can coupling constants be determined by visual inspection?

A203 (b). Here $\nu_{AB}/J_{AB} = 0$. (c) is similar to case (a) in frame 202. (a) gives no information about J_{AB} but as long as $\nu_{AB} \neq 0$ there will always be side components.

F205 The exact chemical shifts of the two protons in the distorted cases can be derived from the positions of the components using a simple mathematical relationship.

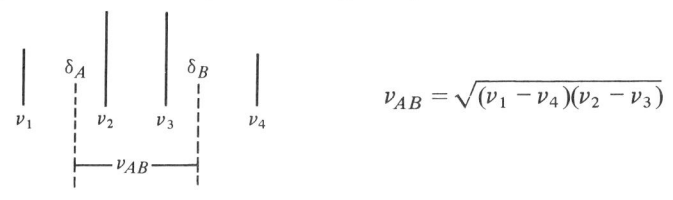

$$\nu_{AB} = \sqrt{(\nu_1 - \nu_4)(\nu_2 - \nu_3)}$$

Calculate ν_{AB} from the spectrum given in Fig. 4.7 and use the δ-scale to derive δ_A and δ_B.

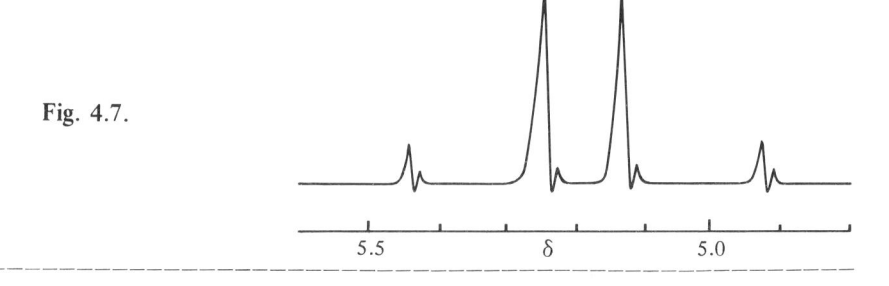

Fig. 4.7.

A204 All except (e). The coupling constant remains unaffected by chemical shift differences.

F206 The cases of the two coupled protons outlined above are relatively simple and can often be easily recognized. Case (c) of frame 202 is, however, very similar to a 1:3:3:1 quartet produced by coupling with three equivalent protons. Careful measurement of the splittings is often sufficient to distinguish between them. Which of the multiplets shown in Fig. 4.8 corresponds to the splitting by three identical protons?

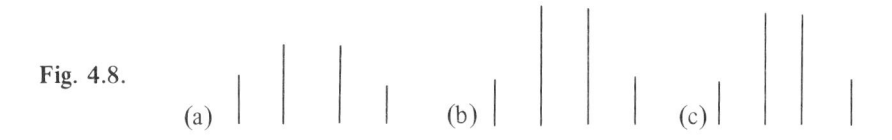

Fig. 4.8.

(a)   (b)   (c)

A205 $\nu_{AB} = 0.25$ p.p.m.; $\delta_A = 5.32$ p.p.m.; $\delta_B = 5.08$ p.p.m.
δ_A and δ_B are obtained by adding and subtracting $\nu_{AB}/2$ from the centre of the quartet.

F207 A notation has been developed to describe the various situations (a)–(e) illustrated in frame 202, in which the alphabet is used to characterize the protons. Case (a) is designated AX and labels one proton A and the other X, the widely separated letters of the alphabet signifying a large chemical shift difference ($\delta_A - \delta_B$) compared with the magnitude of the coupling constant J. When the chemical shift is greater than ten times the magnitude of the coupling constant, ($\nu_{AB}/J_{AB} = 10$), distortion of the intensities is very small. In those cases in which the chemical shift is less than ten times the magnitude of the coupling constant, ($\nu_{AB}/J_{AB} < 10$), the distortion is pronounced and becomes increasingly so as the ν_{AB}/J_{AB} ratio decreases. In these cases the symbols AB are used. Three of the five cases in frame 202 can be described as AB; which are they?

A206 (b). The spaces between the components of a quartet resulting from splitting by three identical protons will always be the same. (a) and (c) do not meet this requirement.

F208 We have introduced a nomenclature for coupling between single protons. We can extend this nomenclature to other systems by including subscripts to denote the number of equivalent protons where appropriate.

Acetaldehyde, CH_3CHO, in which the chemical shift difference is very much larger than the coupling constant would be denoted A_3X. What notation would you employ to describe the following molecules?

(a) CH_3CH_2- of ethanol (spectrum in frame 125);

(b) $\underset{H}{\overset{Cl}{>}}C=C\underset{Br}{\overset{H}{<}}$ (c) $\underset{CH_3}{\overset{CH_3}{>}}CHNO_2$

A207 (b), (c) and (d). Note that (e) is represented as A_2.

F209 In cases where three sets of protons are coupled together and the chemical shifts are large compared with the coupling constants, the symbols $A_aM_mX_x$ are employed. If only two of the three sets of protons are widely separated the notation is $A_aB_bX_x$. If none are widely separated, $A_aB_bC_c$ is employed. Thus, 1-nitropropane:

$$CH_3-CH_2-CH_2-NO_2$$
$$1.0 \quad 2.1 \quad 4.25 \text{ p.p.m.}$$

is designated $A_3M_2X_2$.

Suggest notations for the following examples:

(a) $\begin{array}{c}-O \\ H\end{array} C=C \begin{array}{c} H \\ H \end{array}$ (the spectrum is in frame 189).

(b) $\begin{array}{c} CH_3 \\ CH_3 \end{array} CH-C \begin{array}{c} O \\ H \end{array}$ (make use of Chart 1).

A208 (a) A_3X_2 (b) AB (c) A_6X

F210 We have referred to protons as being equivalent (or non-equivalent) without specifying whether the equivalence is chemical and/or magnetic in origin. This is because in the examples we have chosen the two factors have been indistinguishable. However, chemically equivalent protons are not always magnetically equivalent. For instance, *para*-disubstituted benzenes (4) have two pairs of chemically but not magnetically equivalent protons.

In order to distinguish between these magnetically inequivalent nuclei a prime is used and this system is described as $AA'BB'$. The two protons A and A' are chemically equivalent and exhibit identical chemical shifts. The magnetic non-equivalence arises because the coupling with proton B is *ortho* for A and *para* for A', and with proton B' is *para* for A and *ortho* for A'. Magnetic equivalence requires that J_{AB} should equal $J_{AB'}$ which is clearly not the case here. Group together those parameters in the following list which have identical values in a *para*-disubstituted benzene:

$$\delta_A, \delta_{A'}, \delta_B, J_{AB'}, J_{AB}, J_{A'B'}, J_{A'B}.$$

A209 (a) ABX; (b) A_6MX (large shifts expected).

F211 The n.m.r. spectra of 1,2-disubstituted ethanes, $R^1CH_2CH_2R^2$, may approximate to A_2B_2 or A_2X_2 depending on R^1 and R^2. Some typical spectra for this system are shown in Fig. 4.9. Which alphabetical nomenclature would you use to describe each one?

Fig. 4.9.

(a)

(b)

(c)

A210 δ_A and $\delta_{A'}$; J_{AB} and $J_{A'B'}$; $J_{AB'}$ and $J_{A'B}$.

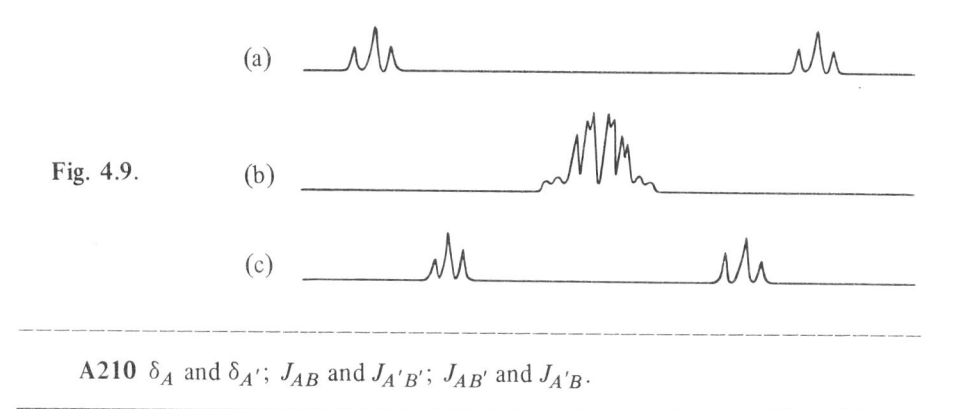

F212 The spectrum of *trans*-1-chloropropene (**5**) can be described as ABX_3 where $J_{AB} \stackrel{\frown}{=} 17$ Hz, $J_{AX} \stackrel{\frown}{=} 1$ Hz, and $J_{BX} \stackrel{\frown}{=} 5$ Hz. In the absence of the X_3 coupling the AB part of the spectrum would be the expected symmetrical 'quartet' which is easily analysed (frames 205 and 206). The introduction of X_3 coupling will often remove the symmetry by superimposing quartets of unequal splitting on the A and B components. Complete the AB part of spectrum (b) in Fig. 4.10.

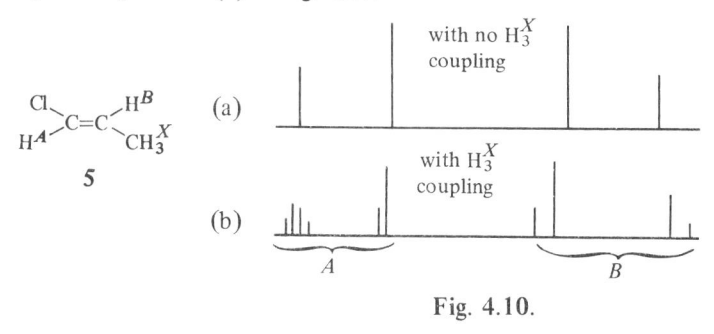

$$\underset{H^A}{\overset{Cl}{\diagdown}}C=C\underset{CH_3^X}{\overset{H^B}{\diagup}}$$

5

Fig. 4.10.

A211 (a) and (c) – A_2X_2; (b) – A_2B_2. Note that the distortion in (c) is very small.

F213 We have now dealt with the notation which is available for the description of the spin states of a wide range of systems. You have already seen the distortion that occurs as the simplest system AX becomes AB. However, as A_aX_x or $A_aM_mX_x$ systems tend towards A_aB_b or $A_aB_bC_c$, then not only do the first-order patterns become distorted but they gain extra lines. Such cases lead to very complex spectra. It is important to appreciate that your ability to analyse coupling patterns visually will therefore be restricted initially to A_aX_x and $A_aM_mX_x$ (first-order) type systems. Certain second-order spectra can, however, be readily recognized by virtue of their symmetry. For example, we have already seen the AB system as a distorted doublet of doublets (often referred to as the 'AB quartet').

$$\underset{X}{\overset{H^A}{\diagdown}}C=C\underset{Y}{\overset{H^B}{\diagup}}$$

6

Again frame 211 showed a typical A_2B_2 system as:

$$\text{X-CH}_2^A\text{-CH}_2^B\text{-Y}$$
7

$AA'BB'$ systems found, for example, in *para*-disubstituted benzenes also give rise to symmetrical patterns. Which of the partial spectra in Fig. 4.11 (all benzenoid absorptions) are produced by *para*-disubstituted benzenes?

(a)　　　　　　　　　(b)　　　　　　　　　(c)

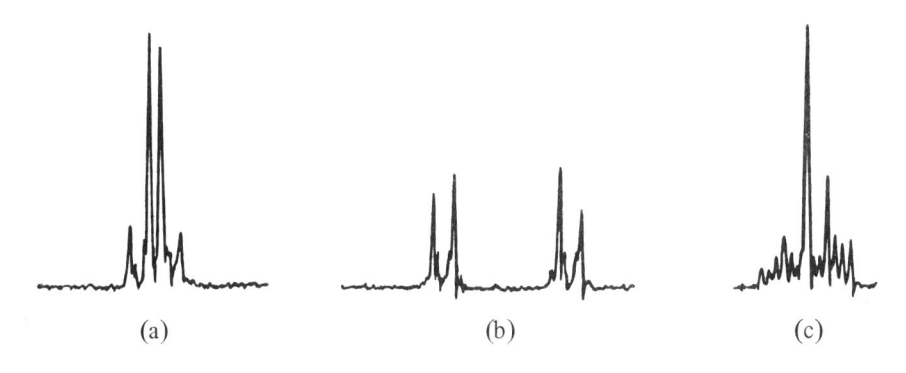

Fig. 4.11.

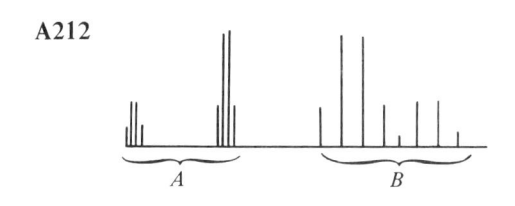

A212

Note that the AB part of the spectrum can still be analysed using the equation in frame 205 once the X_3 coupling has been accounted for. This is only true for ABX_n cases and **not** ABC_n cases.

F214 Which of the following statements is/are false?

(a) $AA'BB'$ refers to the case of chemically equivalent but magnetically non-equivalent protons.
(b) An A_2X_2 pattern is the same as an AX pattern except that each multiplet is twice as intense.
(c) The alphabetical notation does **not** allow the group A_aC_c.
(d) The notation A_2XY is equivalent to ABX_2.

　　A213 (a) and (b). (c) is not symmetrical.

　　A214 (b) A_2X_2 would give two triplets, AX would give two doublets; A_aC_c is the same as A_aB_b.

Aids to the Interpretation of Complex Spectra
F215 (i) High field strengths
We have seen that complete identification of a compound by its n.m.r. spectrum or the complete analysis of a spectrum of a known compound is frequently frustrated because the information from visual inspection of coupling patterns is inadequate. Fortunately methods are available which enable simplification of complex spectra to a degree where extra information can be obtained.

One cause of complexity in spectra relates to the situation where a coupling constant is comparable with the chemical shift difference between the coupled protons (the A_aB_b situation). Now the coupling constant is a parameter of the molecule over which we have no control. By contrast the chemical shift (measured in hertz from some standard) can be altered by changes in an instrumental parameter. Which instrumental parameter?

F216 The chemical shift from TMS in hertz of the methyl protons of ethanol is *larger/smaller* in a 100 MHz spectrum than a 60 MHz spectrum.

    A215 The strength of the applied field.

F217 The chemical shift difference in hertz between the methyl signal and the methylene signal of ethanol is *smaller/larger* in a 100 MHz spectrum than in a 60 MHz spectrum.

    A216 larger. Chemical shift in hertz is directly proportional to the applied field (*cf.* frame 66).

F218 An *AX/AB* coupling pattern in a 60 MHz spectrum tends to become an *AX/AB* pattern in a 220 MHz spectrum.

    A217 larger.

F219 Thus we are able to simplify spectra to a certain extent by using an instrument operating at high field strength. In addition to reducing an *AB* type distorted signal to the more recognizable *AX* form, the use of high fields can also simply separate two signals of two sets of protons which are not coupled, but whose chemical shifts are very similar. The 60, 100 and 220 MHz spectra of 4-bromobutyronitrile, $BrCH_2CH_2CH_2CN$ are shown in Fig. 4.12. The 100 MHz is Fig. 4.12(a). Which of the other two is the 220 MHz spectrum?

    A218 *AB*; *AX*.

F220 (ii) Double resonance
Another common factor which leads to complex spectra is the presence of protons coupled to two or more different protons, for example systems *ABC*, A_2BX and A_3M_2X. Clearly, if it were possible to destroy coupling by one set of protons, say the *X* in A_2BX,

(a)

(b)

(c)

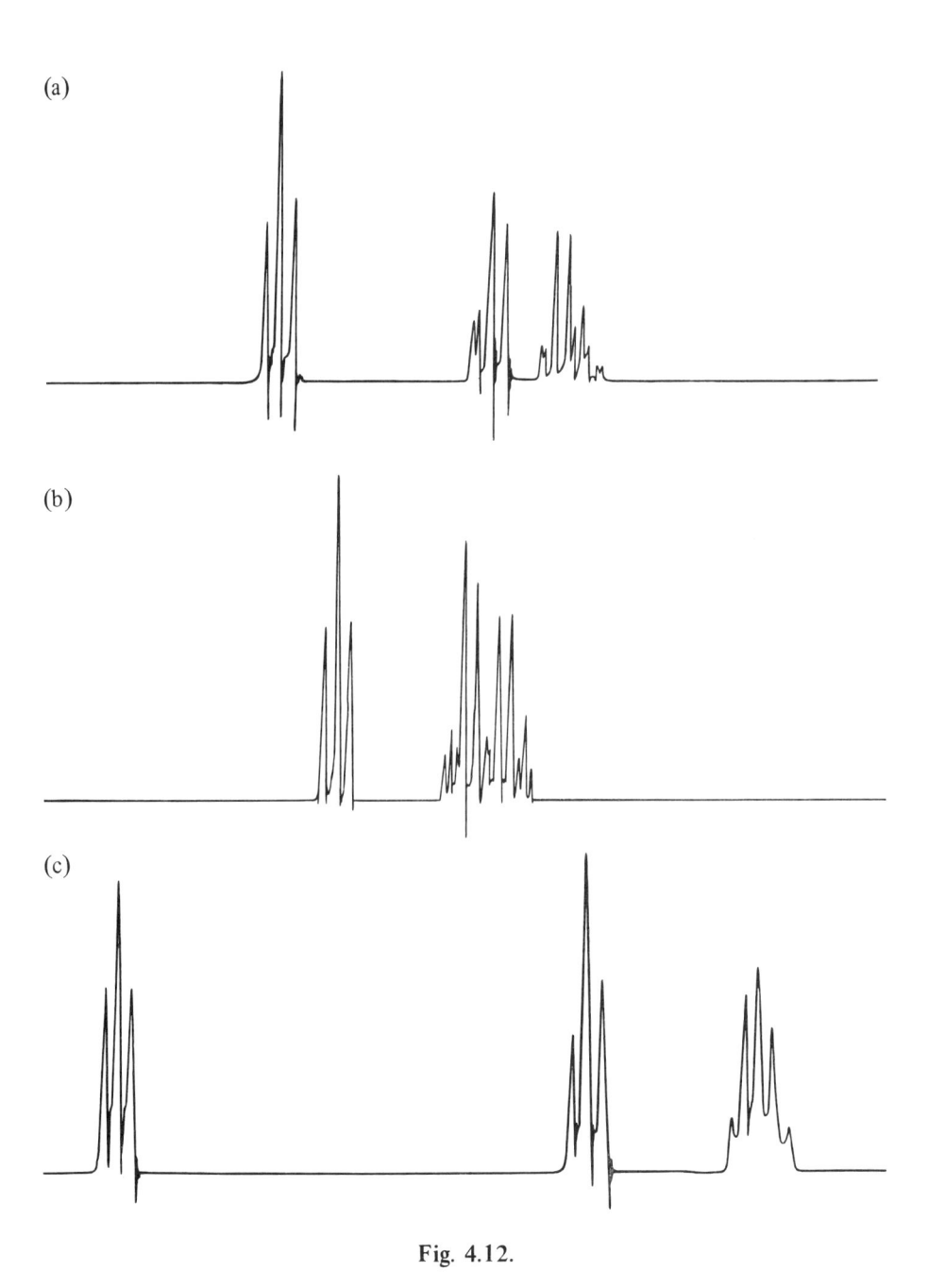

Fig. 4.12.

Reproduced from *Chemistry in Britain* 5(4), p. 150, 1969, Fig. 5 with permission.

there would be a chance of recognizing what is left behind. This can be achieved through the technique of DOUBLE RESONANCE (also referred to as spin decoupling, double irradiation). You recall that spin–spin splitting of the methyl signal in ethanol (spectrum shown in Fig. 4.13) by the methylene protons occurs as a result of the three possible spin states which the two methylene protons can adopt.

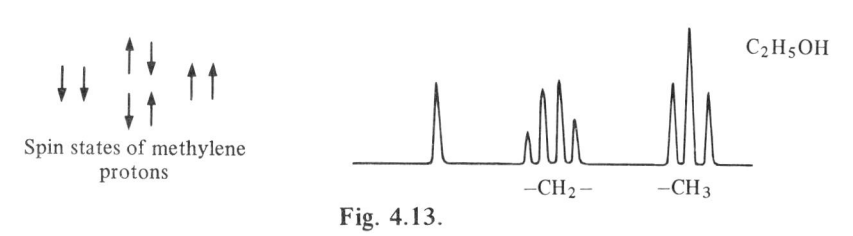

Fig. 4.13.

If now the methylene protons were induced to change their spin state very rapidly, the methyl protons would see only an average 'CH$_2$ state', and coupling would disappear. Such a circumstance obtains when the methylene protons are irradiated strongly with an additional radio frequency at their resonance position. When a spectrum is run under these conditions the methylene absorption is eliminated and the methyl signal is reduced to a singlet (Fig. 4.14).

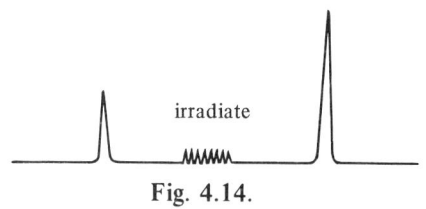

Fig. 4.14.

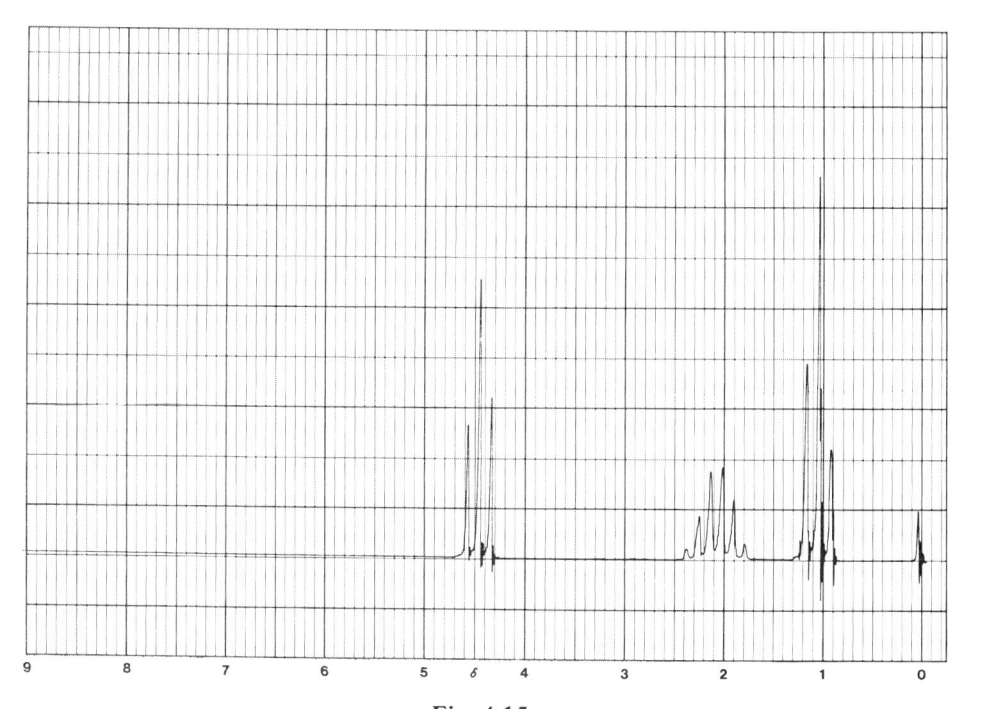

Fig. 4.15.

The spectrum of 1-nitropropane is shown in Fig. 4.15.

Draw diagrammatic representations of the spectra which would result if the sample was irradiated:

(a) at the CH$_2$–NO$_2$ resonance position;
(b) at the C–CH$_2$–C resonance position;
(c) at the CH$_3$– resonance position.

A219 (c). Chemical shift differences (in hertz) increase with increasing applied magnetic field.

F221 We were able to analyse a complex multiplet in a previous example (frames 189–198), by careful visual inspection of the spectrum as a whole. Let us see how simple the analysis becomes when spin decoupling is used.

$$CH_3^e-CH_2^d-O\underset{H^a}{\overset{}{\diagdown}}C=C\underset{H^b}{\overset{H^c}{\diagup}}$$

8

Fig. 4.16.

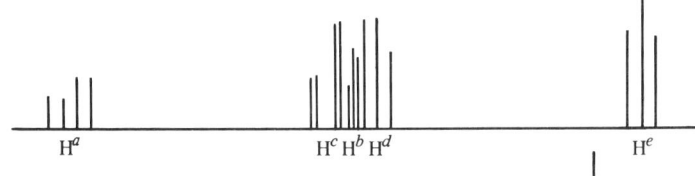

H^a H^c H^b H^d H^e

Consider the overlapping signals of H^b, H^c and H^d (Fig. 4.16). At which resonance position would we have to irradiate to reduce this multiplet to the form shown in Fig. 4.17?

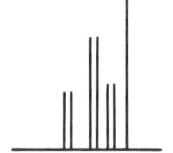

Fig. 4.17.

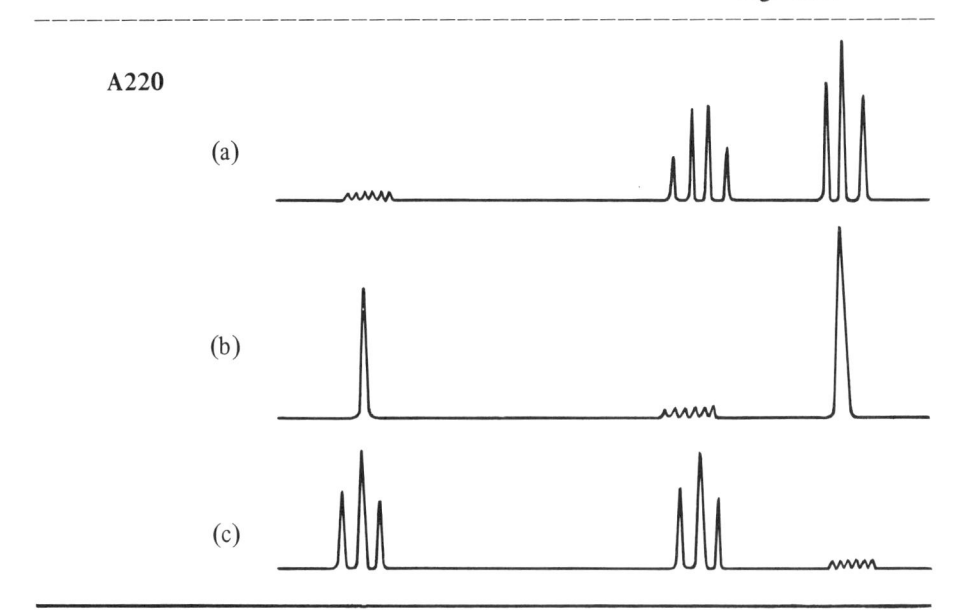

A220

(a)

(b)

(c)

F222 Draw a representation of the central multiplet you would observe if the spectrum was run while the sample was irradiated strongly with a frequency corresponding to the resonance position of H^a.

A221 The methyl resonance position (H^e). Double irradiation has reduced the methylene quartet to a singlet.

F223 Unfortunately, it is not possible to use this technique to remove coupling due to protons whose chemical shift is less than about 20 Hz from the site where simplification is required. Thus it would be difficult to remove coupling between H^b and H^c in the above example.

Which of the following coupled systems would you expect to be capable of simplification by double resonance?

$$ABX, \ A_2B_2X, \ AB, \ A_3X, \ A_2B_3, \ ABC$$

A222

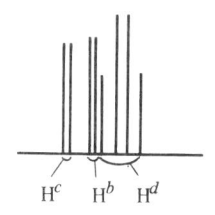

$H^c \quad H^b \quad H^d$

F224 One of the real benefits which results from possessing a spin decoupling accessory with an n.m.r. spectrometer is seen in the case of a spectrum containing several groups of complex patterns. By appropriate choice of double irradiation frequency, the detection of coupled partners becomes relatively easy.

An unknown compound gave spectrum (a) when measured under normal circumstances (see Fig. 4.18). Its spectrum was then re-run three times while the sample was being irradiated separately at resonance B, C and D. The results are shown as spectra (b), (c) and (d) respectively.

Which protons are coupled with the protons responsible for the absorption bands at (i) B, (ii) C, (iii) D, (iv) E?

A223 $ABX, \ A_2B_2X, \ A_3X$. In the other three the chemical shift differences are likely to be too small.

F225 Other analytical information indicated that the compound was an ether of molecular formula C_5H_8O containing one *cis*-disubstituted double bond. What is its structure?

Fig. 4.18(a).

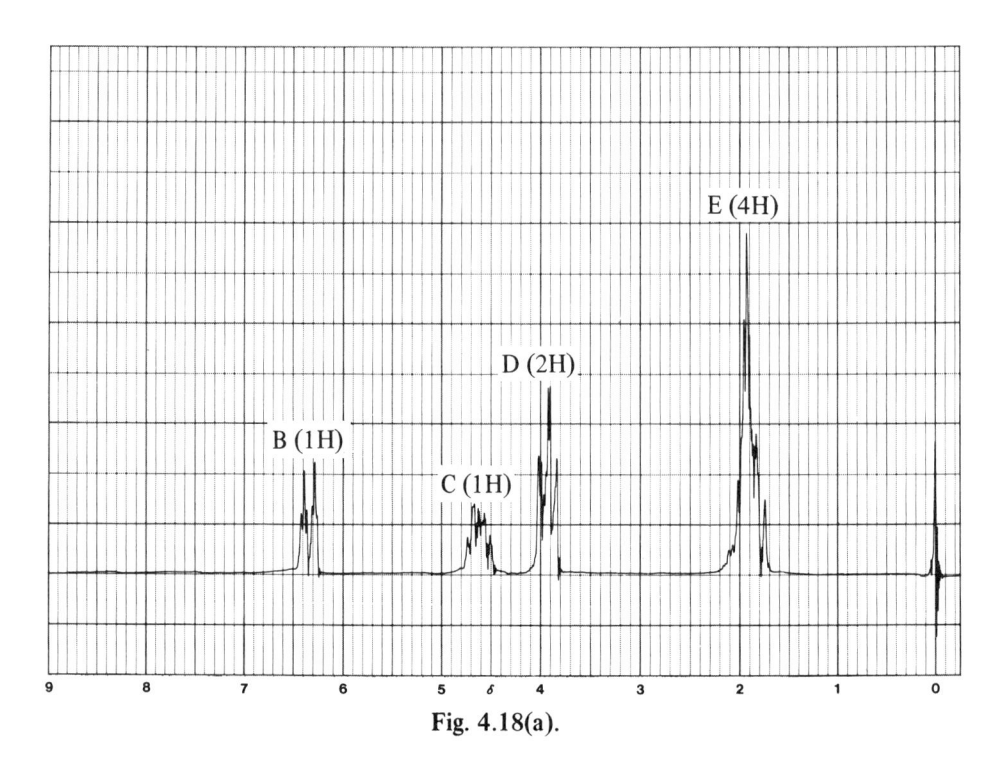

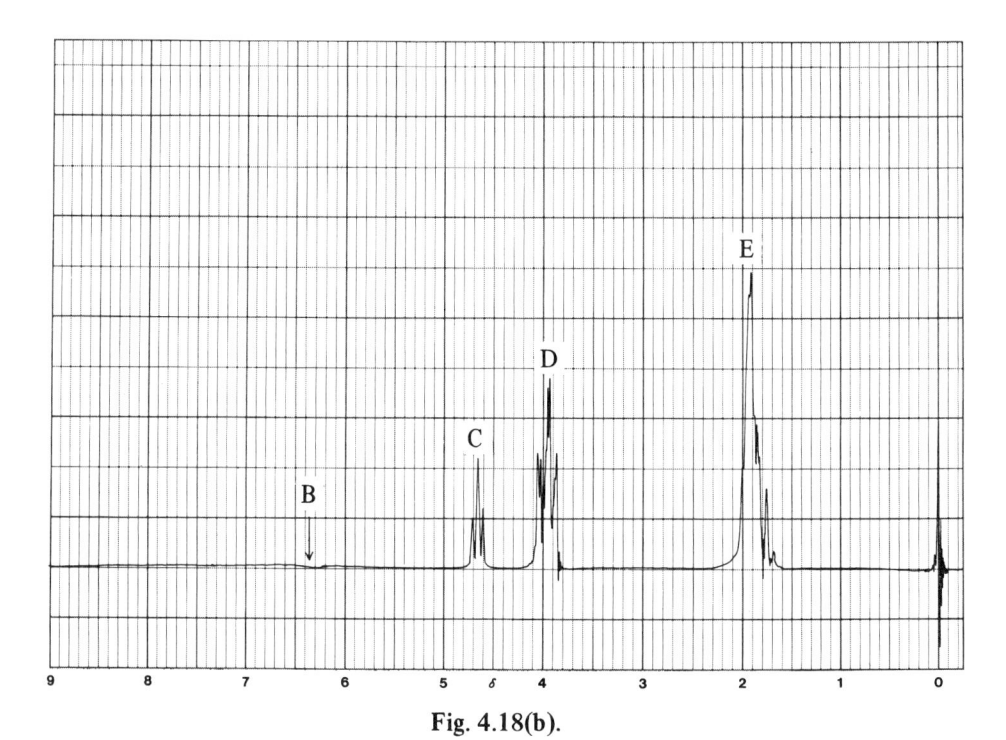

Fig. 4.18(b).

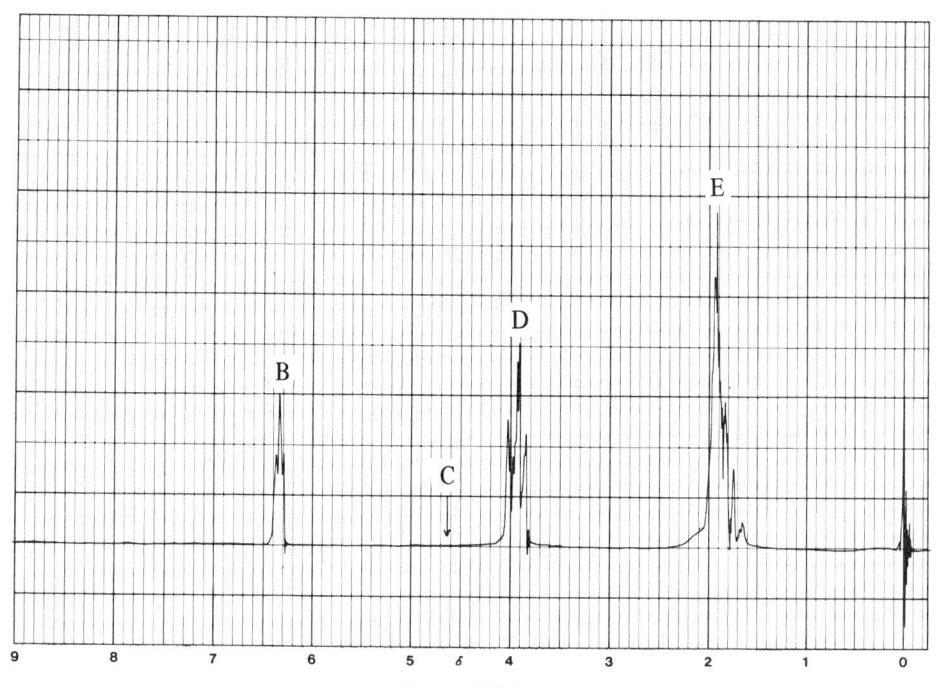

Fig. 4.18(c).

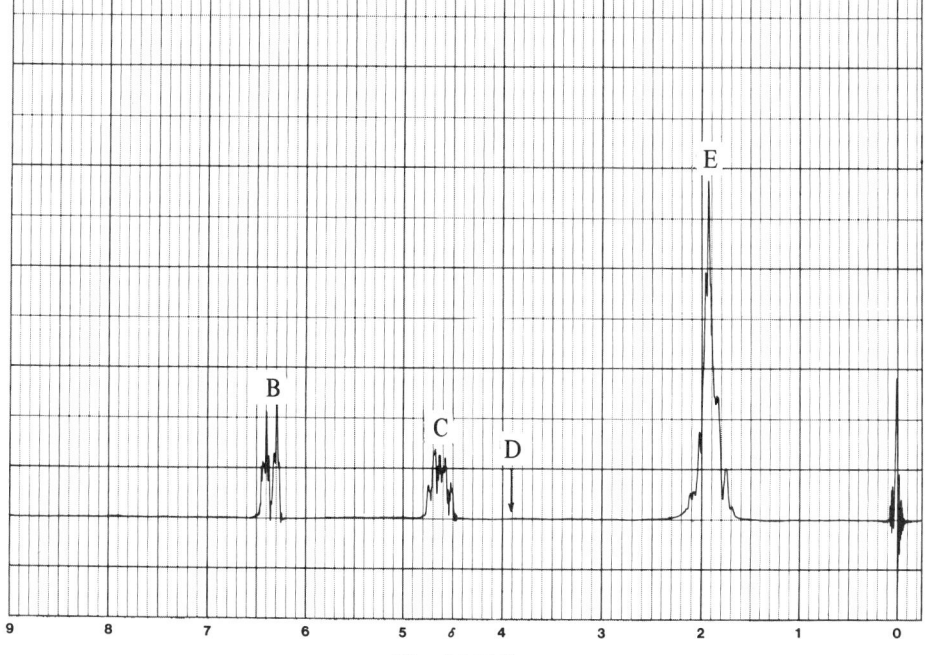

Fig. 4.18(d).

A224 (i) B is coupled to C and E; (ii) C is coupled to B and E; (iii) D is coupled to E; (iv) E is coupled to C, D and B. Spectrum (b) shows that irradiation at B simplifies C but does not reduce it to a singlet. C is therefore coupled to B and is also coupled to either D or E. Spectrum (d) shows that C is unaffected by irradiation at D. Therefore C is coupled to E. As expected B is simplified by irradiation at C (spectrum (c)) but it is not reduced to a singlet. Spectrum (d) establishes that B is not coupled to D and is therefore coupled to E. D is unaffected by irradiation at B or C so its complex structure must arise from coupling with E.

F226 (iii) Deuterium labelling and exchange

The two simplification techniques discussed so far are of limited application. In particular, inherently complex patterns such as A_2B may not be sufficiently simplified towards A_2X for recognition purposes by using a 100 MHz instrument, and a 220 MHz machine may not be available (they are rare at the present time). Furthermore, spin decoupling would be inappropriate in such a case for reasons already mentioned. The following two techniques at least have the potential of solving these problems, but they each have practical drawbacks.

The first concerns the chemical replacement of one of the sets of protons by the hydrogen isotope, deuterium. Although deuterium possesses spin, and hence a magnetic moment, coupling constants with protons are very much smaller than analogous proton–proton coupling constants. Furthermore, resonance of deuterium nuclei would not be observed when a spectrometer is set up to measure protons. Thus, replacement of protons by deuterium removes their signal from the spectrum and also greatly simplifies any coupling they showed with other protons of the system. If the A_2 protons of an A_2BC system were replaced by deuterium, it would be reduced to an $AB/A_2B/AX$ system (which one?) Similarly if the B proton was replaced an $AB/A_2B/AX$ system would result and if the C proton was replaced we would be left with an $AB/A_2B/AX$ system.

A225

B, C, D and E indicate the protons responsible for the B, C, D and E signals in the spectrum.

9

F227 Protons on carbon atoms adjacent to carbonyl, carboalkoxy, or nitro-groups readily undergo base-catalysed deuterium exchange. For instance, the ester $X\overset{\gamma}{C}H_2 \overset{\beta}{C}H_2 \overset{\alpha}{C}H_2 CO_2R$ may be converted into $XCH_2CH_2CD_2CO_2R$ by shaking with $D_2O/NaOD$. The spectrum of the original ester would consist of a triplet for the α-protons, a quintet (assuming equal coupling) for the β-protons, and a triplet for the γ-protons.

How would the spectrum be modified in the deuterated compound?

A226 *AB; A_2B; A_2B.*

F228 The principal limitation of the use of this technique is one of synthesis. Many positions in organic molecules do not lend themselves readily to deuterium replacement. Clearly, this does not apply to exchangeable protons such as these on heteroatoms e.g. OH, SH, NH etc. You recall that these resonate at variable field and their signals are frequently broad. They sometimes turn up buried in a multiplet for example, and occasionally this can cause difficulties in interpretation. By simply shaking the sample of an alcohol with heavy water, D_2O, the original OH resonance can be eliminated with consequent simplification of the spectrum. 2-Mercaptoethanol, with two replaceable hydrogens, readily lends itself to this technique. In the spectrum shown in Fig. 4.19 the hydroxyl proton, which undergoes rapid exchange under normal conditions, appears as a singlet superimposed on one of the components of the low field triplet at 3.6 p.p.m. The SH proton under the same conditions does not exchange rapidly and consequently its signal is split by the adjacent methylene protons and appears as a triplet (1.4 p.p.m.). Furthermore, as the SH proton couples with the adjacent methylene protons, the signal of the latter appears as a complex pattern consisting of an overlapping doublet of triplets.

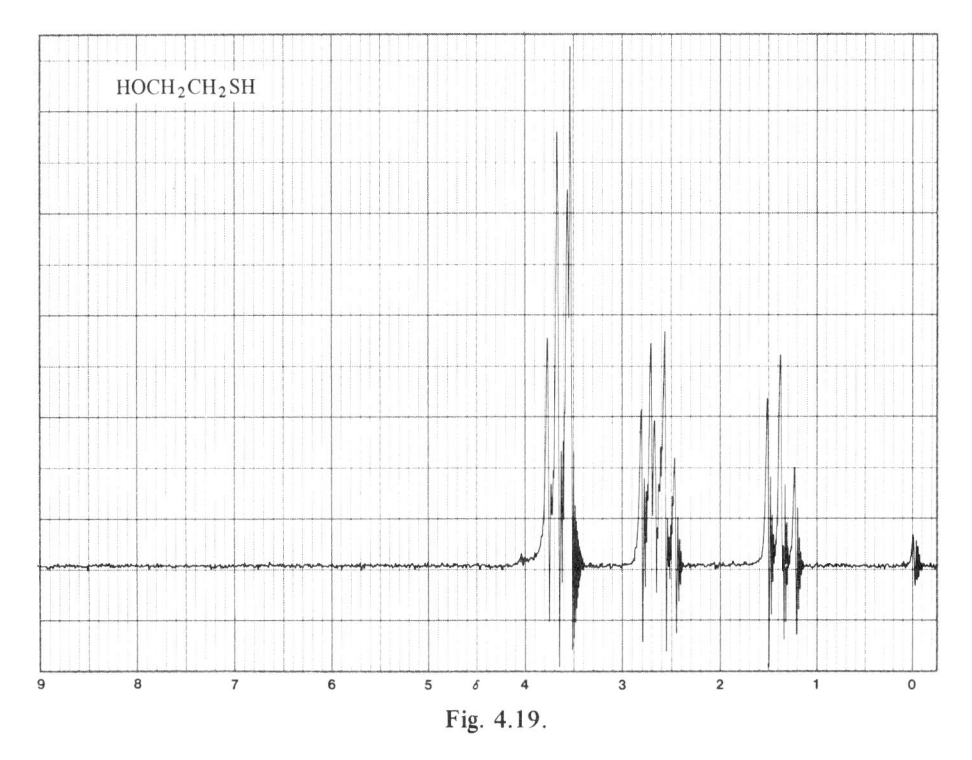

Fig. 4.19.

The whole spectrum may be simplified by the deuterium exchange technique where both SH and OH protons are replaced by deuterium. Draw the spectrum you would expect to obtain after the sample had been shaken with D_2O.

A227 The α-proton absorption would be absent, the β-proton would appear as a triplet, and the signal of the γ-protons would be unaffected.

F229 (iv) Computed spectra

When substitution of hydrogen by deuterium is impractical from a chemical point of view, complex systems may be identified with the use of computed spectra. The mathematical relationships governing coupling patterns are complex but well-known, and theoretical patterns can be calculated given the parameters of chemical shift difference and coupling constant. Many such theoretical patterns, for example AB, AB_2, A_2B_2 and A_2B_3, are now catalogued in the chemical literature. Thus, if one suspects the unknown sample contains, say an A_2B_3 system, one may calculate a series of patterns (or, usually refer to those already calculated) where chemical shift and coupling constant are varied until one is found which fits the experimentally observed signal. Not only does such a fit confirm the original suspicion, but knowledge of the values of the two parameters makes available other useful information.

The multiplet (Fig. 4.20) is an expanded version of the signal of the aromatic protons of a compound suspected of containing a 1,2,3-trisubstituted benzene ring. Furthermore, it is believed that the *meta* substituents are the same, i.e. the compound has the general structure (**10**).

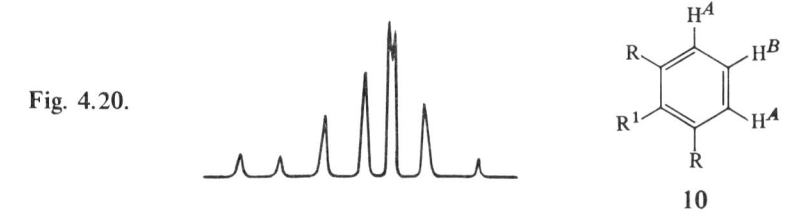

Fig. 4.20.

10

If the structure is correct we are examining an A_2B system, and therefore we should look at tables of computed A_2B spectra for a matching spectrum. Three computed spectra given in Fig. 4.21 illustrate the variation of the multiplet pattern with coupling constant. One of these patterns corresponds to the observed spectrum and therefore not only confirms the A_2B nature of the system but also gives values of the coupling constant and chemical shift difference between the A and B protons. What are these values?

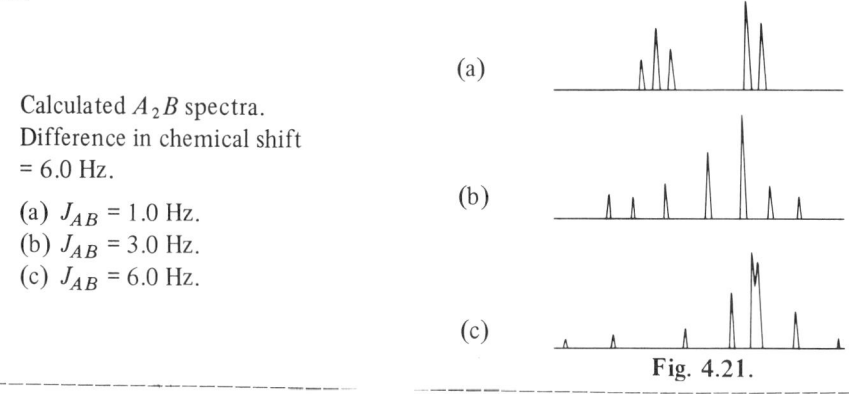

Calculated A_2B spectra.
Difference in chemical shift
= 6.0 Hz.

(a) J_{AB} = 1.0 Hz.
(b) J_{AB} = 3.0 Hz.
(c) J_{AB} = 6.0 Hz.

Fig. 4.21.

A228

```
 5      4      3    δ   2      1
```

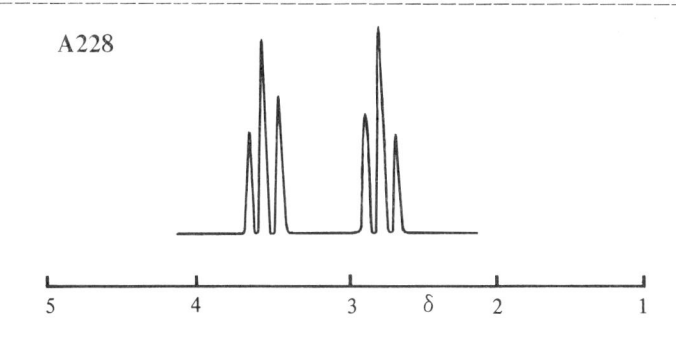

F230 Indicate whether the following statements are true or false.

(a) An A_6X system is so complex that computed patterns are required before it can be recognized.
(b) The main disadvantage of using isotopic substitution for simplifying spectra lies in the chemical problems which may arise in preparing the sample.
(c) If 220 MHz spectrometers were more readily available, their extensive use would render the other three ways of interpreting complex spectra obsolete.
(d) There is little point in using spin decoupling if the coupled partners are much more than 20 Hz apart.
(e) Simplification of an A_3BX_2 system by spin decoupling would be attempted by irradiation of the sample at a frequency equivalent to the X resonance position.
(f) The technique of eliminating resonance due to certain protons by shaking the sample with D_2O is only effective when they are readily exchangeable.

A229 (c) J_{AB} = 6.0 Hz. Chemical shift difference = 6.0 Hz.

F231 (v) Paramagnetic shift reagents

The presence of paramagnetic impurities in a sample usually broadens the absorption bands of the n.m.r. spectrum. This is due to a modification of the relaxation processes in the sample by the unpaired electron of the paramagnetic species.

It has been found, however, that in the case of certain paramagnetic lanthanide complexes, the broadening of the absorption bands is not too severe, and at the same time the bands are displaced from their normal chemical shift values. This paramagnetic shift effect causes a spreading out of the spectrum and often separates absorption bands which normally overlap. The information obtainable from a spectrum can be increased significantly by the use of these paramagnetic shift reagents (or shift reagents as they are sometimes called).

A simple example of the effect of adding a shift reagent to a sample is shown in Figs. 4.22 and 4.23. Figure 4.22 shows the normal spectrum of an aliphatic alcohol dissolved in $CDCl_3$, and Fig. 4.23 a spectrum of the sample after the addition of a paramagnetic shift reagent.

What is the structure of this compound? (Note that the OH absorption band is shifted too far to be observed.)

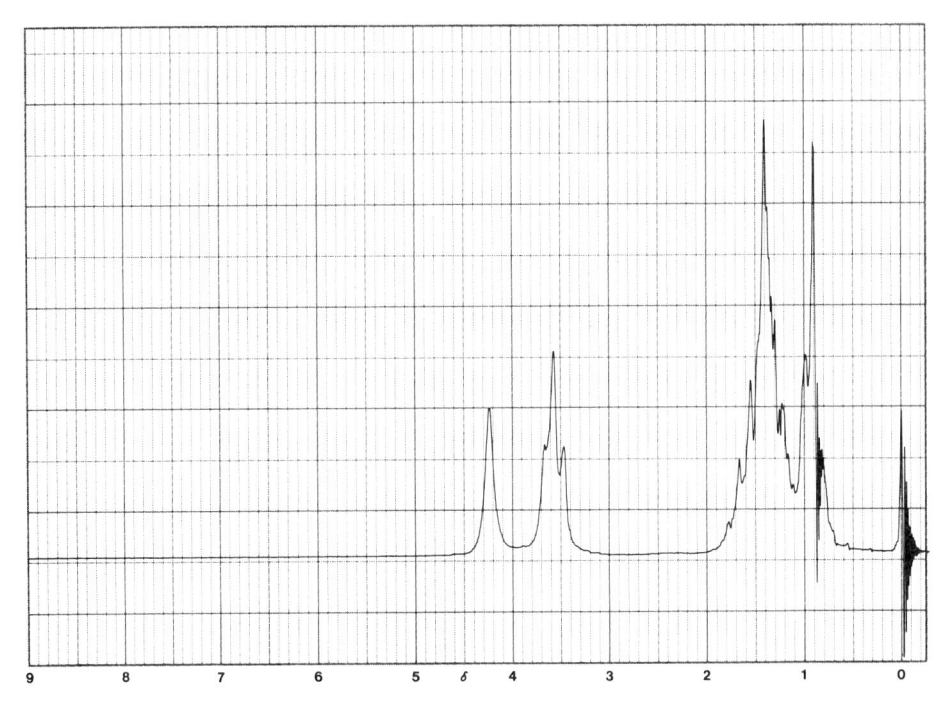

Fig. 4.22.

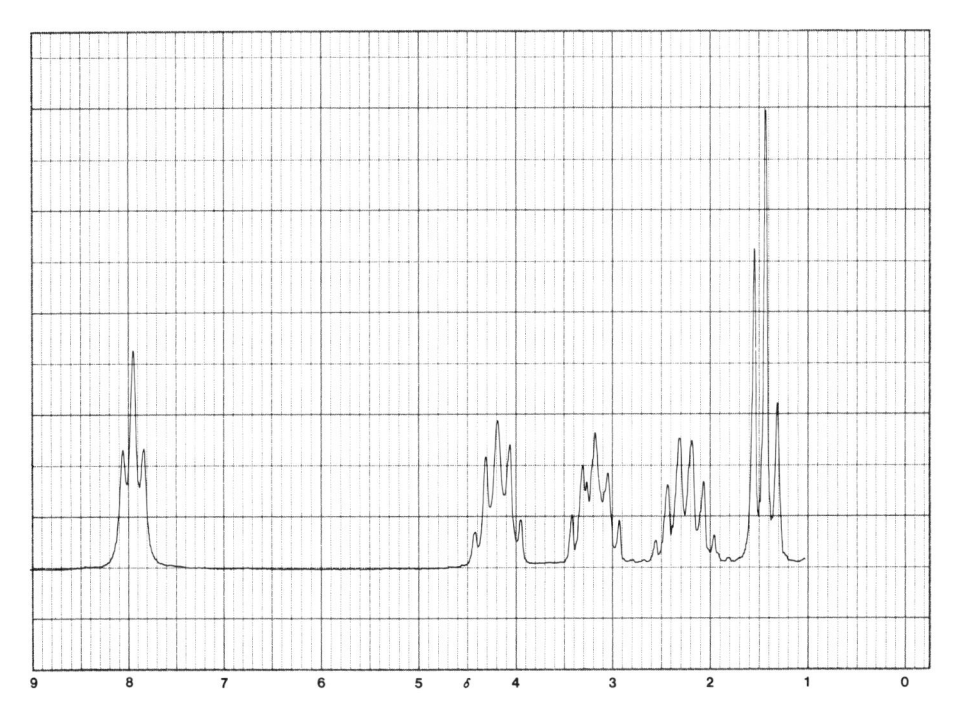

Fig. 4.23.

A230 (a) False. This is a first-order system; A would appear as a doublet and X as a septet. (b) True. (c) False. See frame 226. (d) False. More should be less. (e) True. (f) True.

F232 A widely used chemical shift reagent is tris(2,2,6,6-tetramethylheptane-3,5-dionato) europium [Eu(dpm)$_3$] (11). In complexes of this type, the lanthanide ion can increase its coordination by bonding interaction with lone pair electrons of such groups as NH_2, OH, C=O, $-O-$, CO_2R, CN.

$$\text{Eu}\left[\begin{array}{c} O=C{-}C(CH_3)_3 \\ | \\ CH_2 \\ | \\ O=C{-}C(CH_3)_3 \end{array}\right]_3$$

11

What will be the effect on the spectrum of benzene of adding Eu(dpm)$_3$ to the sample?

A231 HO–CH$_2$–CH$_2$–CH$_2$–CH$_2$–CH$_3$. There are five well resolved absorption bands in the spectrum. The high field triplet is clearly the absorption of a methyl group adjacent to a –CH$_2$– group. The remaining four bands of approximately equal intensity are due to the four –CH$_2$– groups.

F233 It appears that the degree of shift is a function of the distance of the group from the coordination centre.

Which group in the molecule is responsible for the absorption at lowest field (largest δ-value) in the spectrum in Fig. 4.23?

A232 There will be no shift. The benzene molecule has no functional group which possesses lone electron pairs and so there will be no bonding interaction. On the other hand, the presence of this paramagnetic impurity may broaden the absorption band.

F234 The mechanism by which lanthanide samples cause shifts is called the pseudo-contact interaction. It originates from the magnetic field generated by the paramagnetic ion, and as such is most effective in causing shifts in protons close to the site of coordination.

Would you expect, therefore, that in addition to simplifying spectra, paramagnetic shift reagents may also provide information about molecular geometry such as configuration and conformation?

A233 The –CH$_2$– adjacent to the oxygen atom. The hydrogen bonded to the oxygen atom is, of course, shifted the most but as stated in frame 231, its absorption is not shown in Fig. 4.23.

A234 Yes. Predictions of shifts may be made for any molecular geometry and compared with experimental results. There is much research activity in this new subject area but it is already clear that the interaction between the shift reagent and the compound being studied may itself modify the molecular geometry in some cases.

Revision Summary
The alphabetical notation
Spin–spin coupling patterns can be described using an alphabetical notation where protons widely differing in chemical shift are assigned letters widely separated in the alphabet, and those similar in chemical shift are given letters adjacent to one another in the alphabet. Thus we have for example:

CH$_3$CHO	C$_2$H$_5$CO−CH$_2$−CH$_2$−COCH$_3$	CHCl=CH$_2$
12	13	14
AX_3	A_2B_2	ABC

In cases where protons are chemically, but not magnetically, equivalent the prime is used, e.g.

15	16
$AA'XX'$	$AA'BB'$

(15: para-nitroanisole; 16: para-methoxyphenol)

Coupling patterns become more complex as one changes an (AX) system into an (AB) system, i.e. when $\Delta\delta \leqslant J$, spectra can become very complicated.

Interpretation of complex spectra
Useful aids in the interpretation of complicated spectra include:

(i) use of higher field strengths;
(ii) spin-decoupling by double resonance;
(iii) substitution by deuterium;
(iv) use of computed spectra;
(v) paramagnetic shift reagents.

Now try the multiple choice test at the beginning of this part. Write your answers in a vertical column and mark your script as before. Then consult the remarks below.

0–6: Consult your instructor as to whether you should go on or seek extra tuition. If you are using the programme for private study, re-read the frames pertinent to those you answered incorrectly and, when you are satisfied you fully understand the material, read the paragraph at the end of this part and start the next part.

More than 6: Revise the material pertinent to those questions you answered incorrectly. If your instructor does not want you to consult the bibliography or attempt any of the questions which follow, read the paragraph at the end of this part and continue with the next part.

Questions

1. Study the structure and spectrum of compound (17) and draw the spectra you would expect if:

   (i) the sample was irradiated with a frequency equivalent to 1.17 p.p.m.
   (ii) the sample was irradiated with a frequency equivalent to 2.55 p.p.m.
   (iii) H^a and H^b were exchanged for deuterium.

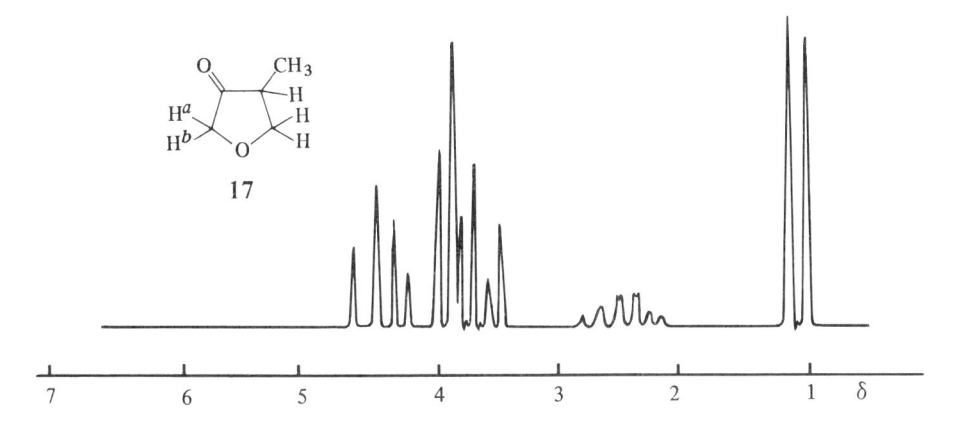

2. Summarize the limitations of the four techniques: deuterium exchange, use of high field strength, double resonance and use of computed spectra, and illustrate the use of each method by an ideally suited example. Use the bibliography to answer this question.

Further Reading

W. McFarlane, *Chem. Brit.* 142 (1969) [double resonance].
J. K. Becconsall and M. C. McIvor, *Chem. Brit.* 147 (1969) [220 MHz n.m.r.].
Y. K. Pan and M. T. Rogers, *Rev. Pure Appl. Chem. (Aust.)* 18, 17 (1968) [survey of computer methods for the interpretation of complex spectra].
B. C. Mayo, *Chem. Soc. Reviews* 2, 49 (1973) [shift reagents].

You are aware of the circumstances which lead to very complex spectra and the ways in which a chemist can simplify matters. You are also in a position to be able to extract most of the structural information available from an n.m.r. spectrum and have already been guided through the interpretation of several spectra. We have provided in Part 6 more opportunities for you to acquire the experience in spectral analysis necessary to improve the effectiveness of the instruction you have received up to now. Before using this, however, you should become familiar with the consequences of the relatively long 'n.m.r. time-scale' and the special topic of geminal magnetic non-equivalence which are discussed in Part 5.

PART 5
TEMPERATURE DEPENDENT SPECTRA AND GEMINAL NON-EQUIVALENCE

Aim
The general aims of Part 5 are: (i) to enable a student to decide when more information about a molecule can be obtained by running a spectrum at temperatures other than room temperature; (ii) to extend his interpretive capacity to include the recognition of the complex features which may be present as a consequence of the molecule being chiral.

New Terms and Concepts
Valence isomerism, energy diagram, time-averaged structures, conformer, Newman projections, accidental equivalence, chirality.

Prior Knowledge
The content of Parts 1 to 4, resonance phenomena, an awareness of the extent to which organic molecules are flexible, geminal methyl groups, asymmetric carbon atom.

Objectives for Part 5
When you have completed this part you should be able to:
1. Recognize structural features in a compound which would cause its n.m.r. spectrum to alter significantly within the temperature range $-100°$ to $+150°C$.
2. Predict the nature of the changes in (1) above.
3. Identify chemically equivalent protons or groups of protons within a molecule which are magnetically different.
4. Identify and interpret simple signals in an n.m.r. spectrum due to chemically equivalent but magnetically non-equivalent protons or groups of protons.

Introduction

You recall that in Part 2 we divided the protons of ethanol into three chemically and magnetically distinct types: methyl, methylene and hydroxyl. The three protons of the methyl group are chemically identical because rotation about the C–C bond is very much faster than any chemical process in which they may become involved. The same is true for the methylene protons. It so happens that the protons within each group in ethanol are also magnetically equivalent because the time scale of n.m.r. absorption is much longer than that of molecular vibrations and rotations at room temperature. In fact, in any dynamic situation where a proton changes its environment faster than the n.m.r. absorption process, an average chemcial shift will be observed and spin–spin coupling simplified. However, there are many processes in organic chemistry whose time scale is of the same order as that of n.m.r., where we can observe both the 'average' and 'frozen' situations by conducting measurement at high and low temperatures, respectively. After reading Part 5, you will be able to recognize those phenomena which are likely to show variable temperature effects of this kind and the type of information that can be obtained by running spectra over a temperature range.

Although the methylene protons of ethanol are chemically and magnetically identical, those of 2-methylbutan-1-ol, $CH_3CH_2CH(CH_3)CH_2OH$, are not at any temperature. The second half of Part 5 will demonstrate why this is so and how the geometric property of chirality can be a source of complexity in spectra. Conversely, once you have understood the origin of geminal magnetic non-equivalence in chiral structures, you will frequently be able to recognize when a molecule contains the feature by a study of its n.m.r. spectrum.

Attempt the following multiple choice test at this point, marking it as described in Part 1. (The correct answer sequence is given on p. 175). Compare your marks with the comments given after the test but do not attempt to seek the correct answers to incorrectly answered questions, unless you score more than 6.

Multiple Choice Test

1. Which of compounds **1, 2** and **3** will have a spectrum which will change significantly as the sample temperature is taken from 0° to 150° C?

2. Consider an experiment where the spectrum of the deuterated derivative of cyclohexane $C_6H_1D_{11}$ is obtained under conditions which eliminate proton–deuterium coupling (double irradiation). Measurements are taken at a series of temperatures between −100° and 0°C. Which of the following changes would you expect to see in the spectrum as the temperature of the sample was increased from the lower to the higher value?

(a) One singlet is replaced by two singlets.
(b) Two singlets are replaced by a multiplet.
(c) Two singlets are replaced by one singlet.

3. The singlet due to the hydroxyl proton in a particular sample of propan-1-ol, $CH_3CH_2CH_2OH$, was observed to broaden as the temperature was lowered. This would be due to the fact that:

(a) the hydroxyl protons were being more restricted in their movement between molecules;
(b) the n.m.r. time scale decreases with decreasing temperatures;
(c) bond rotations in the molecule were being slowed down.

4. Which of the following descriptions would be most appropriate for the appearance of the n.m.r. spectrum of semibullvalene (4) given that the measurement was taken at a temperature where the equilibrium process depicted was too fast for each structure to be 'seen' separately by the instrument?

(a) Five multiplets.
(b) Three multiplets.
(c) One complex multiplet.

5. Which of the following molecules has magnetically non-equivalent methyl groups bound to the same carbon atom?

6. How many methylene groups in the molecule shown (9) contain magnetically different protons?

(a) One.
(b) Two.
(c) Three.

7. Which of the following statements is correct about the series of compounds (**10**) to (**13**)?
    (a) All structures show geminal magnetic non-equivalence.
    (b) Only one structure shows geminal magnetic non-equivalence.
    (c) Only one structure does **not** show geminal magnetic non-equivalence.

10

11

12

13

8. Which of patterns (a) to (c) in Fig. 5.1 would be the **least** likely for the methine proton of molecule (**14**)? Assume the absorption is down-field from that of the methylene group.

(a)

(b)

(c)

$$X-\underset{Y}{\underset{|}{\overset{H}{\overset{|}{C}}}}-CH_2-CO-CH_3$$

14

field ⟶

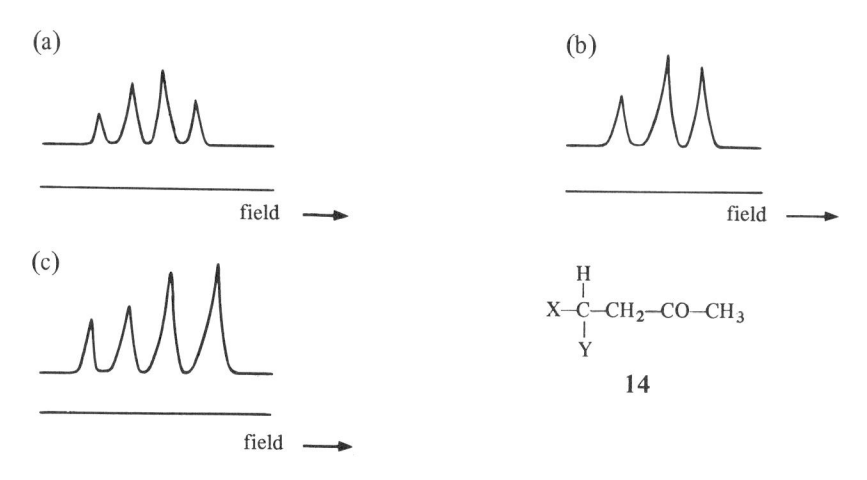

Fig. 5.1.

9. If the two methyl groups of compound **15** were 'accidentally equivalent' their absorption signal would appear as:
    (a) two doublets;
    (b) one doublet;
    (c) a singlet.

15

10. The signals in Fig. 5.2 are all due to a saturated open-chain methylene group. Which one **must** involve geminal magnetic non-equivalence?

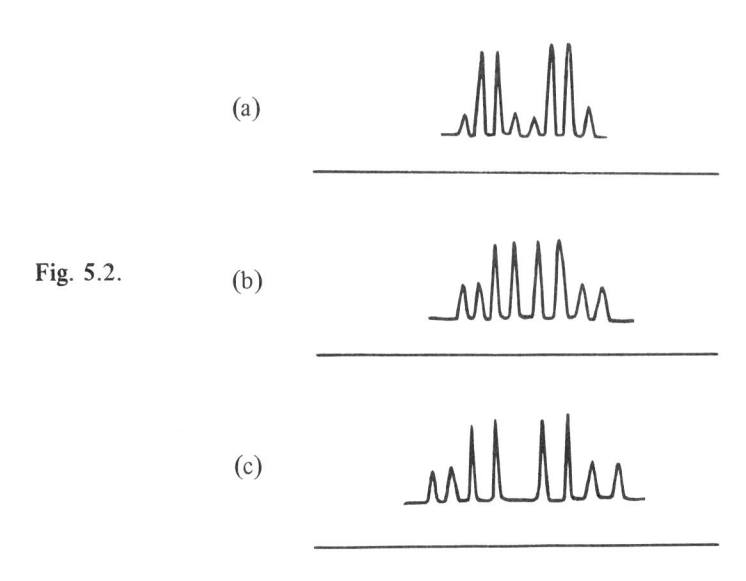

Fig. 5.2.

(a)

(b)

(c)

Comments on marks

Less than 6: Proceed with Part 5.

More than 6: You have at least a superficial knowledge of these subjects already. Check those parts which cover the topics of your incorrectly answered questions and attempt the problems at the end of this part. You are also advised to consult some of the bibliography given there. You should find the problems of Part 6 instructive.

PART 5: TEMPERATURE DEPENDENT SPECTRA AND GEMINAL NON-EQUIVALENCE

Temperature Dependent Spectra

F235 You recall that the resonance signal due to the methyl group in toluene, $C_6H_5CH_3$ appears as a sharp singlet. The three protons are magnetically equivalent because the time scale of nuclear magnetic resonance is slow compared with that of free rotation about carbon–carbon single bond. Thus, in the case of toluene, the methyl group rotates many times during a single absorption process, and all three protons appear identical.

Which of compounds **16** and **17** has a methylene group whose n.m.r. signal appears as a sharp singlet?

16   **17**

F236 An energy barrier must be surmounted before any carbon–carbon bond can rotate. This is usually very small for open-chain carbon–carbon single bonds, and even at the lowest temperatures at which an n.m.r. spectrum can be measured, sufficient thermal energy is acquired from the environment to allow rapid rotation. The barrier against rotation about the carbon–carbon double bond, however, is so high that free rotation does not occur at all, even at the highest temperatures currently available for n.m.r. measurement. Thus we can understand why the methyl protons of ethyl chloride (**18**) are magnetically equivalent whereas the two geminal protons of vinyl chloride (**19**) are different.

18   **19**

Dimethylformamide, $HCON(CH_3)_2$ is an example of a compound containing a partial double bond whose rotational energy barrier is intermediate between the two cases above. The compound is best represented as a resonance hybrid (**20**).

20a   **20b**

20

Its n.m.r. spectrum is found to vary with temperature. Figure 5.3 represents absorption by the methyl groups recorded at 165°C and at room temperature.

Identify the room temperature spectrum.

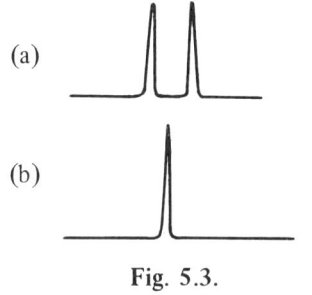

(a)

(b)

Fig. 5.3.

A235 17. Compound **16** is a rigid structure and the two protons of the methylene group are magnetically different, one being on the same side and the other being on the opposite side of the four-membered ring with respect to the OH group.

F237 Modern nuclear magnetic resonance spectrometers are fitted with a variable temperature probe which allows spectra to be recorded at different temperatures, commonly within the range −100° to 200°C. We have seen that the barrier to rotation in dimethylformamide is such as to permit the observation of both free and hindered rotation within this range. In fact, hindered rotation is only one of a number of phenomena common in organic chemistry in which protons change their environment on a time scale suitable for study by variable temperature n.m.r. Inversion in carbocyclic or heterocyclic rings is another. For example, cyclohexane exists as a rapidly equilibrating mixture of conformers, the most stable of which are the two chair forms **21** and **22**, which interconvert by surmounting an energy barrier.

21   **22**

Both H^x and H^y protons change their environment in going from one form to the other. The n.m.r. spectrum of cyclohexane is temperature dependent. At room temperature it consists of a sharp singlet. Does this process have a higher or lower energy barrier than rotation about the C–N bond of dimethylformamide?

A236 (a). At the lower temperature, where rotation about the C–N bond is slow, the two methyl groups will be seen separately, being analogous to the two geminal protons of vinyl chloride. They are in different environments and are expected to appear as two singlets in the spectrum. The availability of greater thermal energy at the higher temperature leads to an increased rate of rotation where the situation resembles that of ethyl chloride, and the two methyl groups become equivalent. Note that at intermediate temperatures the spectrum will appear intermediate between Figs. 5.3(a) and 5.3(b).

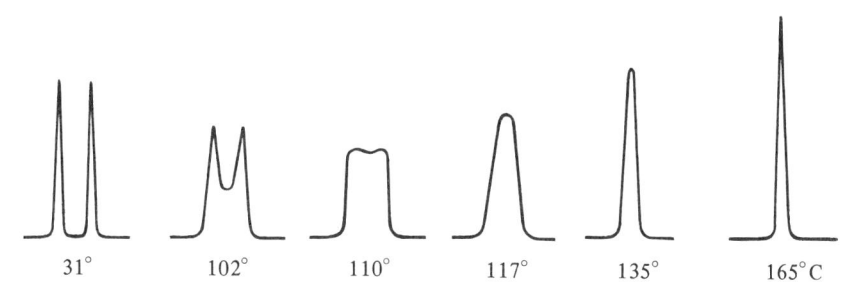

31°   102°   110°   117°   135°   165°C

Beware of the potential misconception that the observation of hindered rotation at low temperatures is a result of a greater contribution of the double bonded canonical form (20b). In a true resonance situation, the relative contribution of canonical forms to the structure is independent of temperature.

F238 Figure 5.4 shows the spectrum of cyclohexane. Would this be measured at:
(a) 100°C;
(b) −100°C?

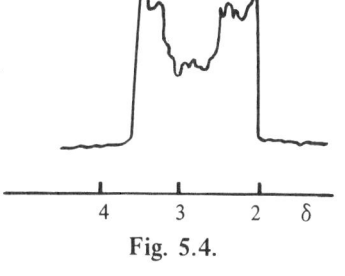

Fig. 5.4.

A237 Lower. You recall that at room temperature the spectrum of dimethylformamide shows an absorption for each methyl group i.e. the n.m.r. spectrometer can distinguish between them because the rate of rotation about the C–N bond is slow. Clearly, at room temperature the interconversion of the conformers of cyclohexane is still proceeding too rapidly for the spectrometer to 'see' any one of them. Therefore less energy is required to surmount the barrier than in the case of dimethylformamide.

F239 Non bond-breaking intramolecular motions are not the only phenomena whereby protons change their environment on this time scale. The spectrum of methanol recorded at room temperature does not usually show spin–spin splitting and consists of two singlets (Fig. 5.5).

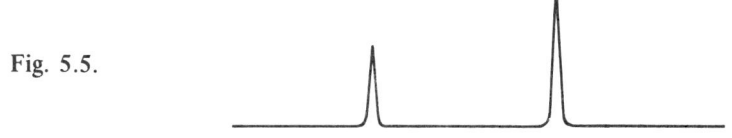

Fig. 5.5.

Think back to an earlier part of the programme and explain why this is so.

A238 −100°C. The sample must be cooled so that the energy available for interconversion is reduced and it proceeds sufficiently slowly for each conformer to exist long enough to be 'seen' by n.m.r. The absorption shows partially resolved axial and equatorial protons, each coupled to the other and to those on adjacent carbon atoms.

F240 If the rate of exchange of the hydroxylic protons can be reduced drastically, then coupling would be observed. Proton exchange, like any other process where bonds are made and broken, proceeds via an energy barrier. Consequently, its rate will diminish as the temperature is lowered. When the spectrum of methanol is recorded at low temperatures (e.g. −40°C), coupling can be observed. Draw the spectrum you would expect under these conditions and at an intermediate temperature, say −5°C.

A239 Hydroxylic protons exchange rapidly, and no single proton stays long enough on an oxygen atom to 'see' the separate spin states of the neighbouring methyl groups. Furthermore, this means that the methyl protons cannot 'see' the separate spin states of the hydroxyl proton.

F241 Valence isomerization is another phenomenon which involves a change in environment of the protons of a molecule by a bond cleavage process, and which frequently occurs at a rate suitable for study by variable temperature n.m.r. Consider the compound barbaralone (**23a** and **23b**).

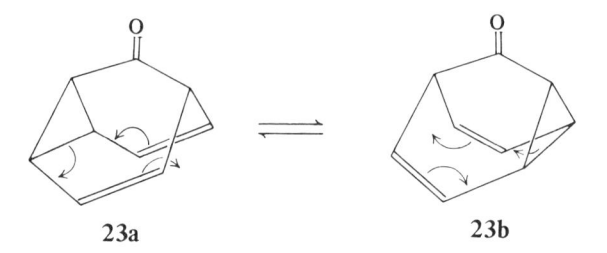

Each molecule can be converted into a chemically identical form by a synchronous bond-breaking and bond-making action proceeding through a symmetrical low energy transition state (**23c**) (Fig. 5.6).

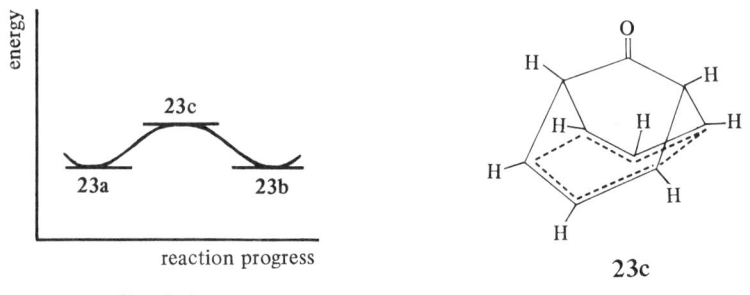

Fig. 5.6.

Below −90°C, the n.m.r. spectrum one observes corresponds to the frozen structure **23a** (or **23b**). At −30°C, however, the protons are observed at positions corresponding to a time-averaged structure between **23a** and **23b**. Which of the following descriptions would be most appropriate to the n.m.r. spectrum of barbaralone at room temperature?

(a) Three groups of multiplets of intensity ratio 1:2:1.
(b) Four groups of multiplets of equal intensity.
(c) Two groups of multiplets of equal intensity.

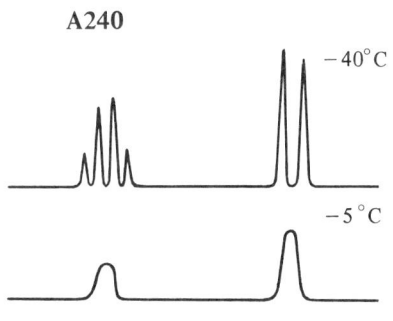

A240

The first spectrum is the normal splitting pattern one may expect for an AX_3 system. The second shows broad bands intermediate between the singlets observed at high exchange rate and the multiplets at low exchange rate.

F242 Semibullvalene (**25**) and dihydrobullvalene (**26**) both contain a bonding system similar to that in barbaralone (**24**).

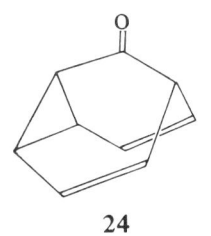

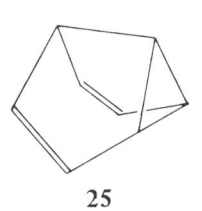

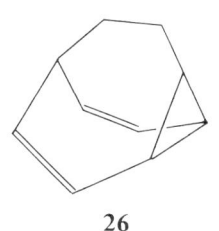

The n.m.r. spectrum of **25** shows three multiplets at room temperature, and this remains unchanged right down to −110°C. Dihydrobullvalene (**26**) shows a 'time-averaged' spectrum at room temperature, but this begins to collapse at 0° and at −90°C the spectrum of the 'frozen' molecule is recorded.

Recalling that the n.m.r. spectrum of **24** at −90°C corresponds to a 'frozen' structure and at −30°C a time-averaged structure, arrange barbaralone (**24**), semibullvalene (**25**) and dihydrobullvalene (**26**) in order of increasing energy barrier for the valence isomerization.

A241 (a). The time-averaged structure of barbaralone resembles the transition state **23c** where one would predict three 'different' types of protons H^a (2), H^b (4), H^c (2).

F243 In summary, there are many processes in chemistry whose rates put them within the scope of study by variable temperature n.m.r. and they include:

(i) rate processes involving reversible intermolecular proton transfer;
(ii) rotations about single bonds with partial double bond character;
(iii) inversion of carbocyclic and heterocyclic rings;
(iv) valence isomerizations and intramolecular rearrangements.

Write down one example for each of the categories (i) to (iv).

A242 Semibullvalene, barbaralone, dihydrobullvalene. Both dihydrobullvalene and barbaralone show 'frozen molecule' spectra below −90°C whereas semibullvalene shows a time-averaged spectrum right down to −110°C. The rate of isomerization of semibullvalene is therefore considerable even at this temperature and the reaction must have a very low energy barrier.

Barbaralone shows a time-averaged spectrum at −30°C, whereas the spectrum of dihydrobullvalene at this temperature is already beginning to collapse towards the 'frozen' structure. Thus at −30°C barbaralone is isomerizing at a faster rate than dihydrobullvalene.

F244 We have made a rather superficial qualitative survey of the scope of variable temperature n.m.r. studies. Their real value, however, comes from quantitative treatment which, although beyond the scope of this programme, should be mentioned briefly at this point. It is possible to deduce the rate of change in proton environment by a detailed analysis of signal shapes. Thus rates at various temperatures can be obtained for these processes and knowledge of the dependence of the rate of a reaction on temperature leads directly (through thermodynamics) to energies of activation.

Continue with the next section.

A243 (i) Proton exchange in methanol.
(ii) Dimethylformamide.
(iii) Inversion of cyclohexane.
(iv) Barbaralone, semibullvalene, dihydrobullvalene.

Geminal Non-Equivalence

F245 At ordinary temperatures, the two methylene protons of ethanol have the same chemical shift. This is because any geometrical arrangement of the atoms (conformer) can be converted into its mirror image by rotation about the C–C bond (demonstrated below using projections of the molecule along the C–C bond).

27a   rotation of CH$_3$ clockwise by 2θ   27b

In other words, mere rotation of the C–C bond can place one methylene proton in an identical environment to that experienced previously by the other, and vice versa. Furthermore, we know that such rotation occurs very rapidly at ordinary temperatures. Consider now ethylene chlorohydrin, ClCH$_2$CH$_2$OH, in the conformation shown in **28a**. Is it possible to convert this into its mirror image by C–C bond rotation?

F246 Again, rapid rotation means that we have two sets of equivalent protons. Is it possible to convert a conformer of R$_1$R$_2$CHCH$_2$OH into its mirror image by rotation about the C–C bond?

A245 Yes.

rotation of the CH$_2$Cl group clockwise through $120° - 2\theta$

28a → 28b

F247 In other words, the two methylene protons in a compound of this kind can never be in identical environments. They are thus chemically and magnetically different, and can have different chemical shifts. Would this hold at all temperatures?

A246 No. Rotation of the –CHR$_1$R$_2$ group of **29a** can never lead to **29b**.

29a   mirror images   29b

F248 This will always happen where methylene groups are attached to carbon atoms which are bound to three different groups (asymmetric carbon atoms). You recall from basic stereochemistry that the presence of an asymmetric carbon atom in a molecule generally means that the latter will not be superimposable on its mirror image; that is, there will be left-handed and right-handed forms. ('Handedness' is now more frequently referred to as chirality.) However, a molecule need not possess an asymmetric carbon atom to be chiral. The sole requirement is that it should not be superimposable on its mirror image. The protons of –CH$_2$– groups in open-chain systems will be magnetically non-equivalent when the molecule containing them is chiral. In practice, the commonest form of chirality is that which involves an asymmetric carbon atom as the chiral centre. Which of compounds **30** to **34** contains –CH$_2$– groups, the protons of which are not magnetically equivalent? That is, which molecules show geminal non-equivalence? (Study the structures carefully. Chirality is not the only feature responsible for geminal non-equivalence!)

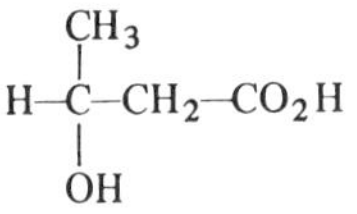

30 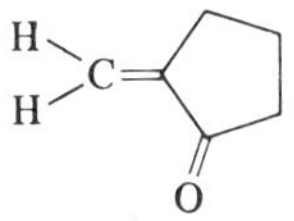   31

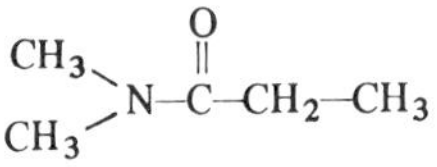

32 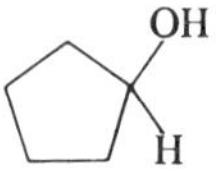   33 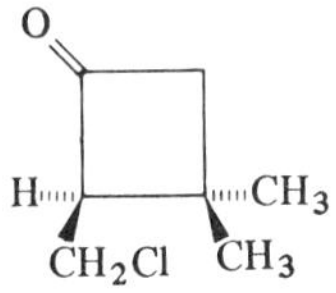   34

A247 Yes. In such cases, a particular conformer's mirror image could not be formed without breaking and reforming bonds.

F249 It follows from frame 248 that the methylene group need not be bound directly to the asymmetric carbon atom, neither is this type of non-equivalence confined to two protons. For example, the principle applies to compounds containing *gem*-dimethyl groups (see structure to right).

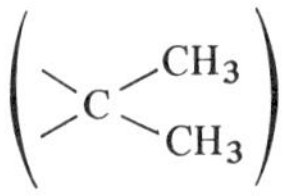

Combining both of these points, we can say that the methyl groups of compound **35** are magnetically non-equivalent, even though the isopropyl group is not directly attached to the asymmetric carbon atom.

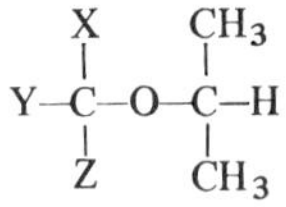

Which of compounds **36** to **41** contains magnetically non-equivalent methyl groups or methylene protons?

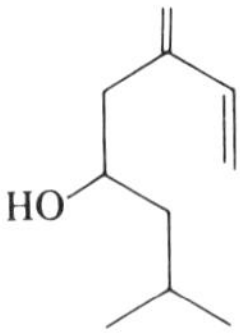

36 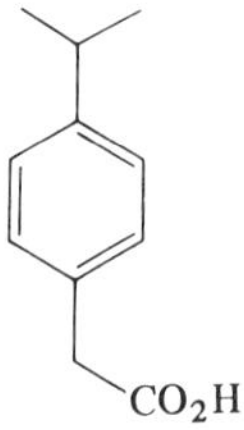   37 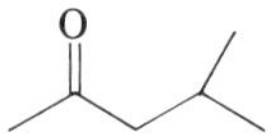   38

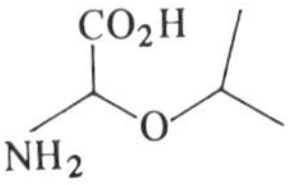

39 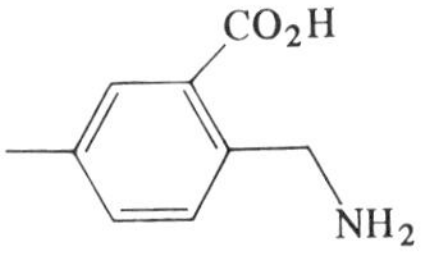   40 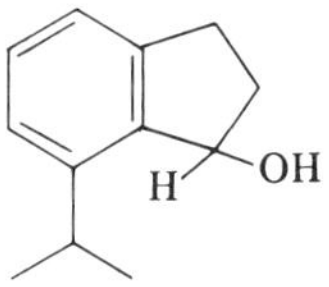   41

A248 Compounds **30, 31, 33, 34**.

$$\begin{array}{c} CH_3 \\ | \\ H-C-CH_2-CO_2H \\ | \\ OH \end{array}$$

30

Chiral centre (an asymmetric carbon atom).

31

No asymmetric carbon atoms and no free rotation. However H^a is in a different environment from H^b.

33

All the CH_2's of the ring have one of their protons on the same side as H_1 and the other on the same side as the OH group.

34

Asymmetric carbon atom also the ring methylene group, see **33**.

F250 Although methylene protons (or *gem*-dimethyl methyl groups) may in principle be magnetically non-equivalent in a particular compound, they may appear equivalent in an n.m.r. spectrum because the chemical shift differences between the partners of each pair is not large enough to be resolved by the instrument, i.e. they are ACCIDENTALLY EQUIVALENT. This is most likely to be the case when the asymmetric carbon atom or other chiral centre is well removed from the methylene group, but it can also occur when it is situated in a neighbouring position. Let us consider the example of diethyl acetyl succinate (**42**). Which carbon atoms (*a–j*) bear hydrogen atoms which are not magnetically equivalent?

$$^aCH_3-^bCH_2-O-\overset{O}{\underset{}{C}}-\overset{^fCH_3}{\underset{H}{\overset{|}{\underset{|}{^dC}}}}-^gCH_2-\overset{O}{\underset{}{^hC}}-O-^iCH_2-^jCH_3$$

42

A249 Compounds **36, 39** and **41**. **37, 38** and **40** do not contain an asymmetric carbon atom.

36

39

41

* Asymmetric carbon.
† Magnetically non-equivalent methylene or methyls through asymmetry.
‡ Magnetically non-equivalent methylene through geometry.

F251 Which of these three is most likely to show magnetic non-equivalence in a routine n.m.r. spectrum?

A250 *b*, *g* and *i*. Once it is established that a molecule is chiral then all the methylene groups will contain non-equivalent protons.

F252 This happens to be the case. A detailed analysis of the n.m.r. spectrum of diethyl acetyl succinate (Fig. 5.7) shows protons on ^bC to be accidentally equivalent and the same applies to those on ^iC. Those of ^gC, however, are seen to be magnetically non-equivalent, having different chemical shifts, *g* and *g'*.

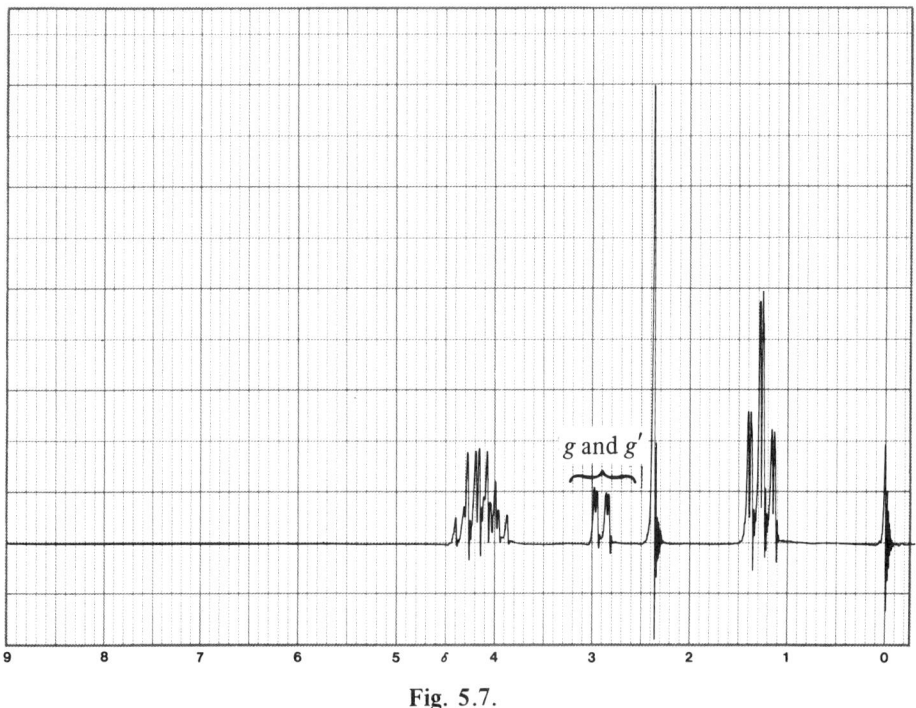

Fig. 5.7.

Let us now examine the structure of 2,2,4-trimethylpentane-1,3-diol (**43**) where the protons have been labelled *a*–*i*. What features of this structure show geminal magnetic non-equivalence?

$$\begin{array}{c} CH_3^a \\ \diagdown \\ CH^c \\ \diagup \\ CH_3^b \end{array} \begin{array}{c} H^d \quad CH_3^f \\ | \quad\quad | \\ -C-C-CH_2^h-OH^i \\ | \quad\quad | \\ OH^e \quad CH_3^g \end{array}$$

43

A251 *g*. This methylene group is closest to the asymmetric carbon atom, ^dC.

F253 The spectrum of this compound, run in the presence of D_2O to remove the OH absorption, is shown in Fig. 5.8.

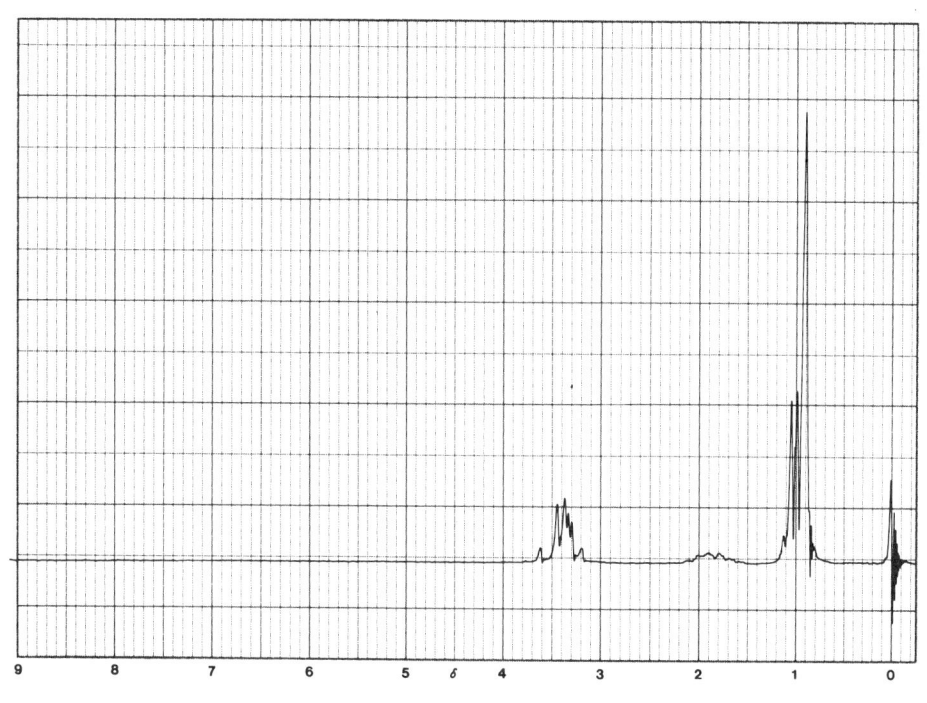

Fig. 5.8.

Which protons are responsible for the multiplet near 1 p.p.m.?

A252 $\left.\begin{array}{c}CH_3^a\\CH_3^b\end{array}\right\} : \left.\begin{array}{c}CH_3^f\\CH_3^g\end{array}\right\} : CH_2^h$

F254 Study the structure of the molecule and predict the shape of the signal for CH_3^a and CH_3^b:
(a) if they were accidentally equivalent;
(b) if they were observably magnetically non-equivalent.

A253 The four methyl groups.

F255 Now predict the shape of the signal of CH_3^f and CH_3^g:
(a) if they were accidentally equivalent;
(b) if they were observably non-equivalent.

A254

(a) [doublet] The doublet arises from coupling with H^c.

(b) [three peaks] Both coupled separately with H^c.

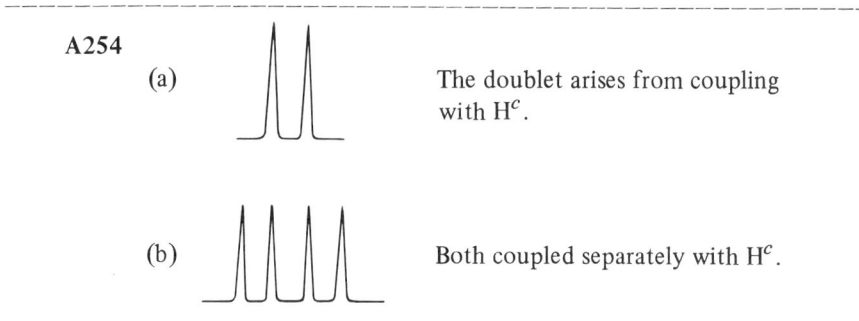

F256 Compare answers 254 and 255 with the actual multiplet (noting particularly the intensities of the components) and deduce which of the following statements is true.
(a) CH_3^f and CH_3^g alone are accidentally equivalent.
(b) CH_3^a and CH_3^b alone are accidentally equivalent.
(c) In both pairs (CH_3^f and CH_3^g : CH_3^a and CH_3^b) the methyl groups are accidentally equivalent.
(d) In neither pair are the methyl groups accidentally equivalent.

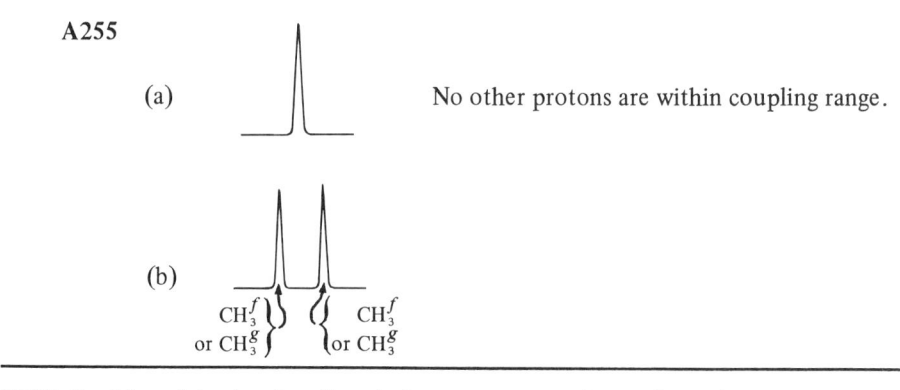

A255

(a) [singlet] No other protons are within coupling range.

(b) [doublet pair] CH_3^f or CH_3^g | CH_3^f or CH_3^g

F257 So although both pairs of methyl groups are equidistant from the asymmetric carbon atom, one pair is accidentally equivalent whereas the other pair can be seen to be magnetically non-equivalent in the n.m.r. The methylene protons H^h are the same number of bonds removed from the asymmetric carbon atom as the methyl protons of CH_3^f and CH_3^g. Let us see if we can deduce whether the two protons of the methylene group are accidentally equivalent or not.

Both the methylene protons CH_2^h and the single proton H^d are bound to carbon atoms bearing hydroxy groups and so both will absorb at a similar low field position between 3.25 and 4.4 p.p.m. (Chart 1). The multiplet near 3.4 p.p.m. must therefore arise from a combination of both

$$\begin{array}{c} CH_3^a \\ {}^{>}CH^c-C-C-CH_2^h-OH^i \\ CH_3^b \quad\quad OH^e\ CH_3^g \end{array}$$
$$H^d\ CH_3^f$$

43

sets of signals and so we must predict the shape of the H^d signal before we can identify the components of the CH_2^h signal. Will H^d give rise to (a) a quartet, (b) a doublet, (c) a pair of doublets?

A256 (a). At first glance, it appears we have a doublet and a singlet, consistent with (c) (answers 254(a) and 255 (a)). But for this to be true, the total intensity of the doublet should be the same as that for the singlet. This is clearly not the case. The doublet is far too weak and therefore must be only half of the two-doublet absorption of answer 254(b), the other half merging with the singlet expected from answer 255(a) to produce an overall singlet three times the intensity of the discernible doublet.

F258 Now look at the methylene protons in structure **43**. How would you describe their expected n.m.r. absorption if they were accidentally equivalent.

A257 (b). Coupling with H^c.

F259 Draw the pattern you would expect if they were not accidentally equivalent and had a chemical shift difference of about 10 Hz.

A258 A singlet.

F260 So the multiplet at 3.4 p.p.m. should consist of a doublet arising from H^d and another singlet if CH_2^h protons are accidentally equivalent or an AB quartet if CH_2^h protons show magnetic non-equivalence. Examine the multiplet and deduce whether the methylene protons are accidentally equivalent or not.

A259

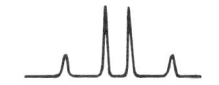

An AB quartet ($\Delta\delta \sim J \sim 10$ Hz)

F261 Are the following sentences true or false?
(a) Magnetic non-equivalence will only be observed in a methylene signal of an n.m.r. spectrum if the chemical shift difference between the two protons is resolvable by the instrument.
(b) Although distance from the asymmetric centre is important, it is not the only factor which determines chemical shift difference between geminal magnetically non-equivalent protons.

A260 They are not.

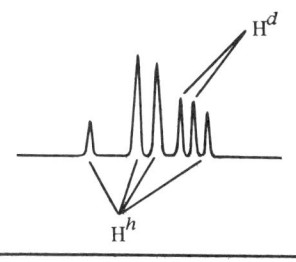

H^d

H^h

F262 We have seen that magnetic non-equivalence of methylene protons can complicate spectra changing, for example, a singlet in the case of compound **44** into an *AB* quartet in the case of compound **45**.

$$\begin{array}{c} H^a \;\; X \\ | \;\; | \\ R-O-C-C-X \\ | \;\; | \\ H^b \;\; Y \end{array} \qquad \begin{array}{c} H^a \;\; X \\ | \;\; | \\ R-O-C-C-Y \\ | \;\; | \\ H^b \;\; Z \end{array}$$

44 45

Further complications may occur if X, Y or Z is a proton (as in compound **46**).

$$\begin{array}{c} H^a \;\; H^c \\ | \;\; | \\ R-O-C-C-X \\ | \;\; | \\ H^b \;\; Y \end{array}$$

46

Now, in general, $J_{ac} \neq J_{bc}$ and this can lead to very complex spectra. An unusually simple example is afforded by the spectrum of 1,2-dibromo-1-phenylethane (**47**) shown in Fig. 5.9. Which of the three signals is due to the methine proton H^c?

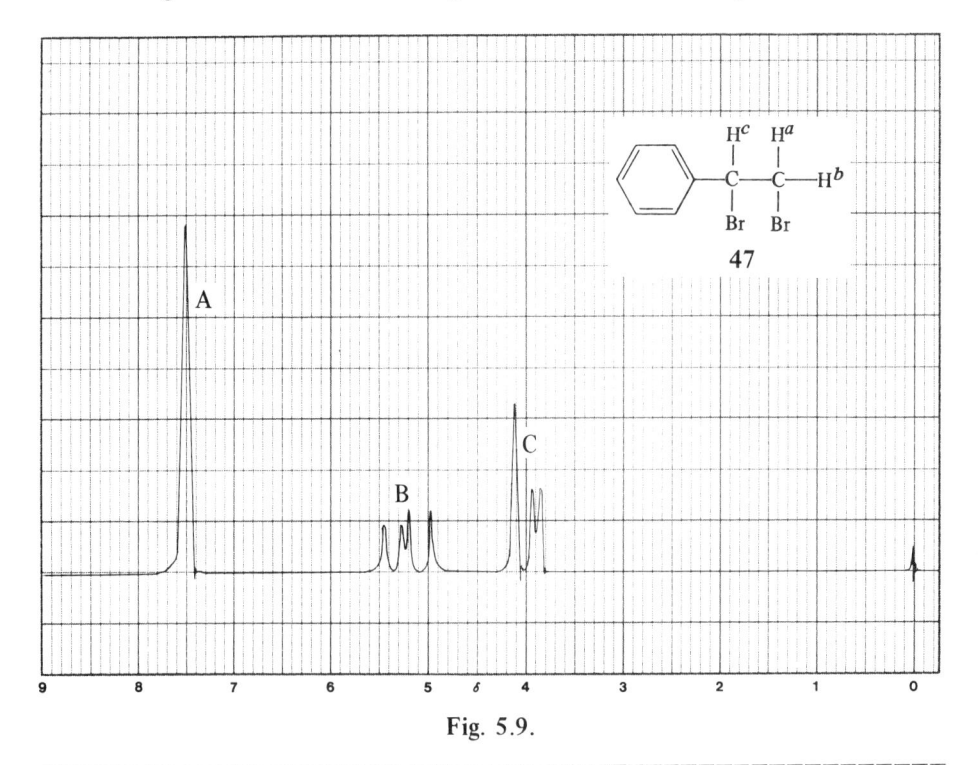

Fig. 5.9.

A261 (a) True; (b) true.

F263 The coupling pattern for this signal is readily interpreted if one assumes $J_{ac} \neq J_{bc}$. Draw a branching diagram to show how it arises.

A262 B. The neighbouring phenyl group accounts for its high δ-value.

F264 Let us turn now to the absorption due to H^a and H^b (signal C) and, for the moment, ignore coupling of these protons to H^c. We do not know the chemical shift difference (if any) between H^a and H^b. Recall how the shapes of the signals due to two coupled protons change as the chemical shift difference between them diminishes (Fig. 5.10).

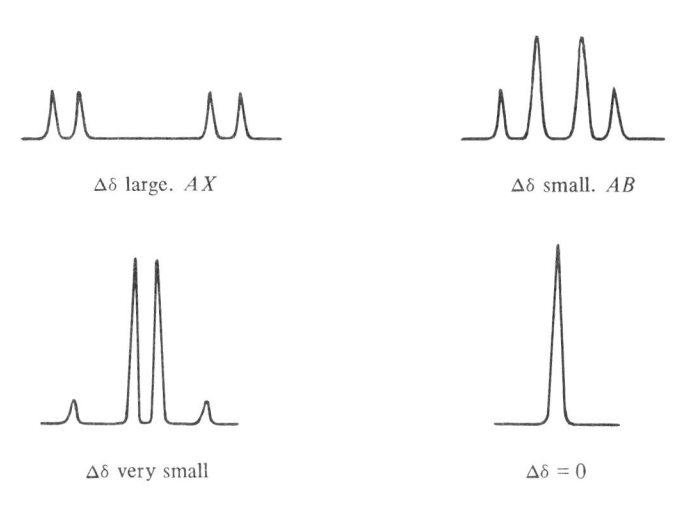

Δδ large. *AX*

Δδ small. *AB*

Δδ very small

Δδ = 0

Fig. 5.10.

For two methylene protons Δδ is likely to be small, very small or zero (accidentally equivalent). Let us assume in this case that Δδ is very small and ignore the weak outer component absorptions of the above theoretical pattern. Furthermore, let us arbitrarily assign the high field component to H^a. Thus, if no coupling to H^c was involved our methylene pattern would appear as shown (Fig. 5.11).

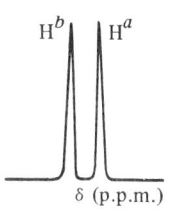

H^b H^a

δ (p.p.m.)

Fig. 5.11.

Now consider the effects of coupling to H^c. Using the branching technique draw the pattern which would result if:

(a) $J_{ac} > J_{bc}$;
(b) $J_{bc} > J_{ac}$.

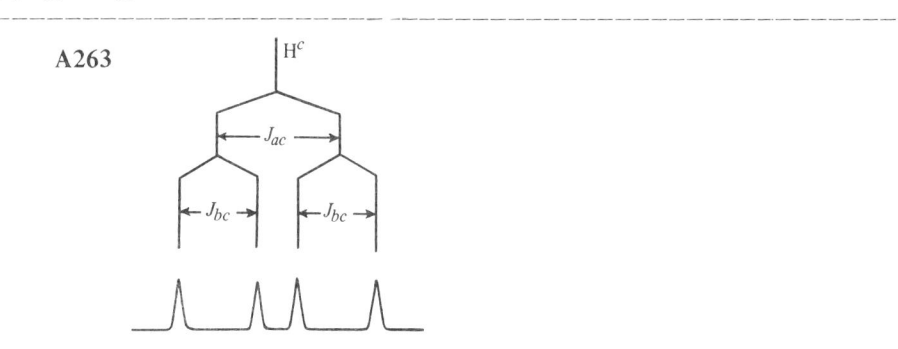

A263

H^c

J_{ac}

J_{bc} J_{bc}

F265 Which of the options (a) and (b) in the previous frame is appropriate for signal C of Fig. 5.9?

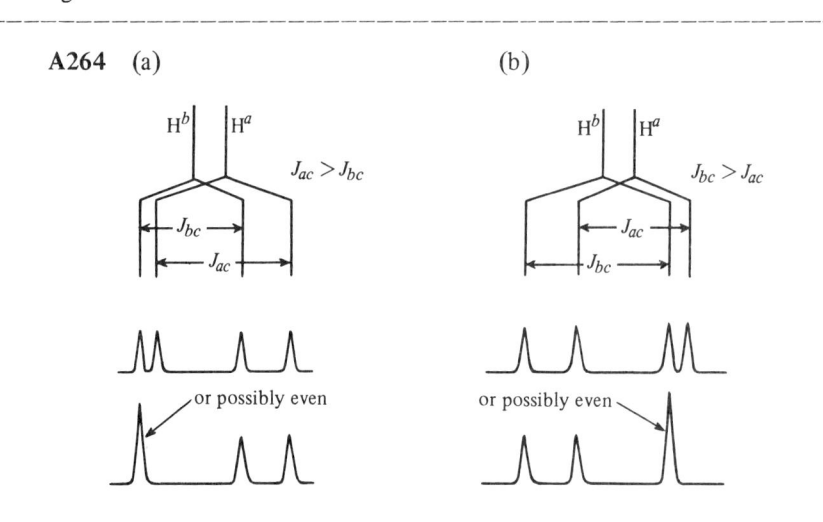

A264 (a)   (b)

H^b H^a $J_{ac} > J_{bc}$ H^b H^a $J_{bc} > J_{ac}$

J_{bc} J_{ac}

J_{ac} J_{bc}

or possibly even   or possibly even

F266 Let us look at one more example. When phenylalanine (48) is measured in D_2O, the NH and OH resonances disappear as a result of deuterium hydrogen exchange, and the spectrum obtained is that given in Fig. 5.12.

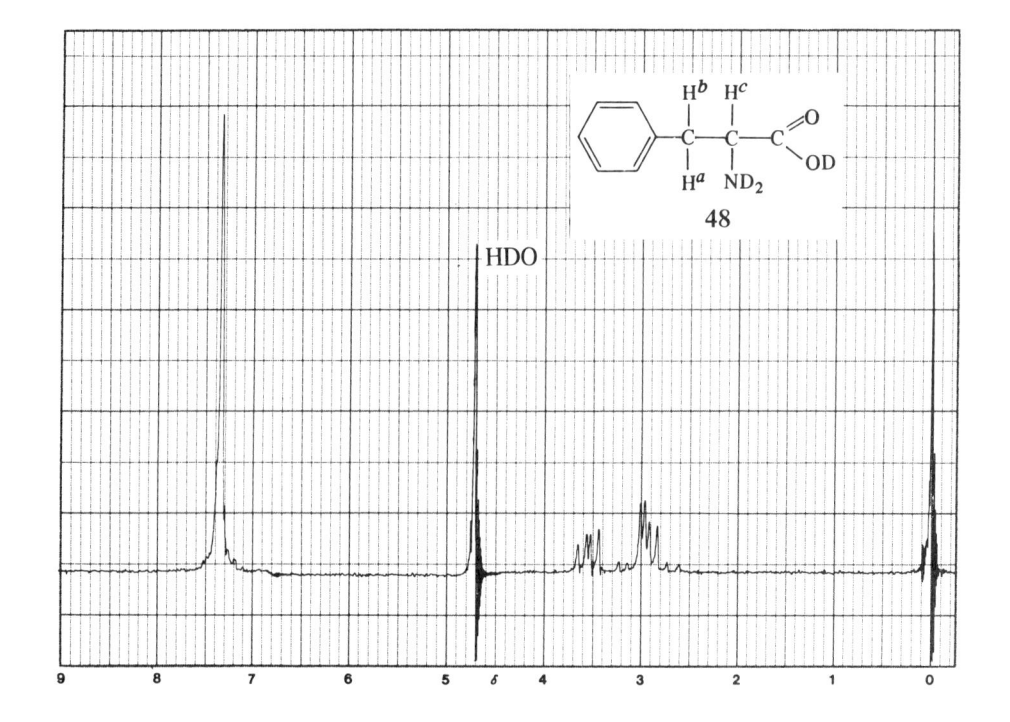

Fig. 5.12.

Ascertain whether or not:

(a) H^a and H^b are accidentally equivalent and $J_{bc} = J_{ac}$;
(b) H^a and H^b have slightly different chemical shifts but $J_{bc} = J_{ac}$;
(c) H^a and H^b have slightly different chemical shifts and $J_{bc} \neq J_{ac}$.

A265 (a).

A266 (c). If (a) applied then the methine signal would be a triplet and the methylene signal a doublet (AX_2). If (b) applied the methine signal would be a simple triplet and the methylene signal two pairs of doublets.

Revision Summary
Variable temperature nuclear magnetic resonance

Molecules in solution are not static but are continually vibrating and rotating as well as undergoing translating motion. The intramolecular motions are responsible for the rapidly changing magnetic environment experienced by most protons in an organic molecule. Generally vibrations and rotations occur more rapidly than the n.m.r. absorption process and so the spectrum of a compound records the average position of each proton. However, there are some phenomena where protons change their environment on a time scale of the same order of magnitude as n.m.r. Such processes are slowed down by lowering the temperature of the sample and speeded up by raising it. Thus, in principle, one is able to see both 'frozen' and 'average' environments by a variable temperature study. Useful information about the process can be gained in this way, and in particular, its energy of activation. The phenomena showing this behaviour include:

(a) restricted rotation, e.g. dimethylformamide;
(b) ring inversion, e.g. cyclohexane;
(c) proton exchange, e.g. methanol;
(d) valence isomerization, e.g. semibullvalene.

Geminal non-equivalence

For the reasons already explained, all the protons of a methyl group are almost invariably magnetically equivalent. The same applies to an sp^3 hybridized methylene group in open-chain compounds **except** when the molecule possesses an asymmetric carbon atom or any other source of chirality. The magnetic difference between the protons in such cases is a fundamental property of the structure and independent of its intramolecular motions. The same would apply to any pair of identical geminal groups, e.g. $-C(CH_3)_2-$.

There are many examples of such groups in organic molecules where the magnetic difference is sufficiently large to be observed in the compound's n.m.r. spectrum. The two protons, in the case of a methylene group, are seen to have different chemical shifts, split each other's signals and have different coupling constants associated with their spin–spin interactions with protons on neighbouring carbon atoms.

When the chemical shift difference is not large enough to be detected by n.m.r. spectra the protons or groups are said to be ACCIDENTALLY EQUIVALENT.

Now have another attempt at the multiple choice test at the beginning of Part 5 marking it in the usual way, and consult the comments which follow.

6 or less: If appropriate, consult your instructor and re-examine those parts of the programme which you did not fully understand until both of you are satisfied your problems have been sorted out. If you are using the programme for private study, read through the sections which led you to incorrect answers and, if necessary, confirm your understanding of the material by consulting the bibliography which follows.

More than 6: Check up on those parts connected with the questions you answered incorrectly and try the problems which follow. You should also spend a little time on the bibliography before proceeding with Part 6.

Questions

1. Explain the following:
   (a) The methyl and methylene resonances of compound **49** are split at low temperatures but collapse to a singlet at high temperatures.

   (b) Two methyl signals are observed in the n.m.r. spectrum of compound **50** at low temperature whereas only one is present at high temperature.
   (c) Two methyl signals are observed in the n.m.r. spectrum of compound **51** at low temperature whereas only one is present at high temperature.

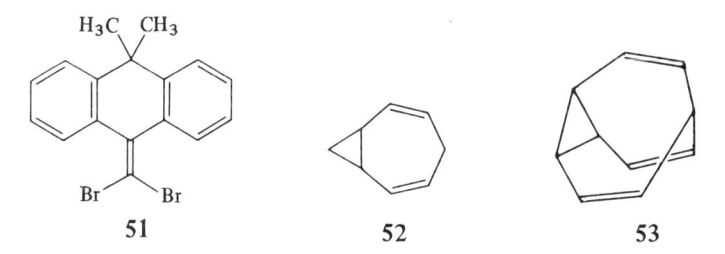

2. Explain the following:
   (a) 3,4-Homotropylidene (**52**) shows four multiplets at 180°C of area ratio 1:2:1:1 whereas at −50°C a much more complex spectrum is observed.
   (b) Bullvalene (**53**) at high temperature has an n.m.r. spectrum consisting of one singlet.

3. Predict the n.m.r. spectrum of the allene (**54**) assuming that the geminal magnetic non-equivalence present would be observable.

4. What is the full structure of the compound Ph–C$_5$H$_{11}$O whose partial n.m.r. spectrum is given below?

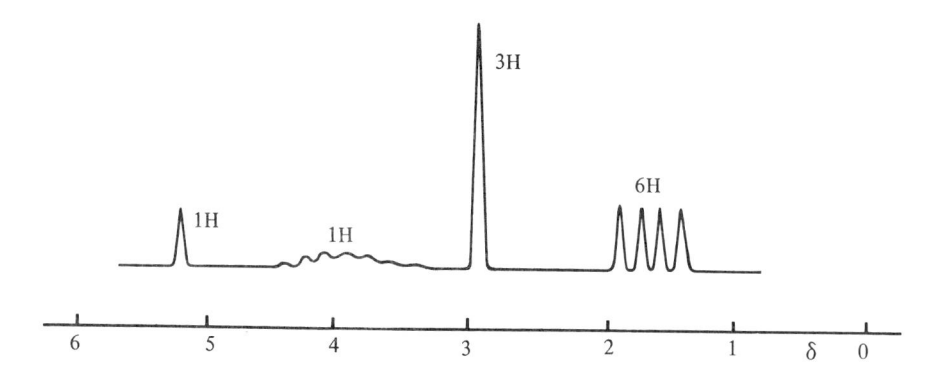

Further Reading

L. M. Jackman and S. Sternhell, *Applications of Nuclear Magnetic Resonance Spectroscopy in Organic Chemistry*, Pergamon Press, Oxford, 1969, Part 5.
J. W. Emsley, J. Feeney and L. H. Sutcliffe, *High Resolution Nuclear Magnetic Resonance Spectroscopy*, Volume 1, Pergamon Press, Oxford, 1965, Chapter 9.
E. D. Becker, *High Resolution Nuclear Magnetic Resonance*, Academic Press, New York, 1969, Chapter 10.
K. Binsch, *Topics in Stereochemistry*, Volume 3 (Eds. N. L. Allinger and E. L. Eliel), John Wiley and Sons, New York, 1968, p. 97.
F. A. Cotton, *Accounts of Chemical Research*, 1968, p. 257; *Chem. Brit.* 345 (1968).
M. Van Gorkam and G. E. Hall, *Quart. Rev.* 14 (1968).
K. Mislow, *Topics in Stereochemistry*, Volume 1 (Eds. N. L. Allinger and E. L. Eliel), John Wiley and Sons, New York, 1968, p. 1.
F. A. Bovey, *Nuclear Magnetic Resonance Spectroscopy*, Academic Press, New York, 1967, Chapter 7.
W. W. Paudler, *Nuclear Magnetic Resonance*, Allyn and Bacon, Boston, 1971, Chapter 8.

You now appreciate why the temperature at which an n.m.r. spectrum is taken can be of the utmost importance in certain kinds of organic compounds. You are also aware of the special kinds of complications that can occur in the spectra of chiral compounds and are able to take this into account in spectral analysis. Having completed the programme this far it only remains for you to gain some further experience in the interpretation of spectra. This is the subject of Part 6.

**PART 6
PROBLEMS**

Objective for Part 6
When you have completed Part 6 you should be able to interpret the n.m.r. spectrum of a known substance or use the n.m.r. spectrum of an unknown substance to formulate plausible alternatives for its structure. You should be able to achieve this more rapidly and effectively than before.

Introduction
'Practice makes perfect'. To finish the programme we have included a part devoted solely to problem-solving exercises. These range from multiframe problems in which each step of the interpretation is spelled out, to single frame exercises requiring you to make your own full spectral analysis. You may have noticed that in earlier parts, we have followed a fairly standard procedure at the beginning of an interpretation, and you will probably find that the following sequence is a useful guide in tackling most problems.

1. Measure the height of each step of the integration curve and add them together. If the molecular formula is known, divide the total integration height by the number of protons and so obtain a value for the integration per proton. Use this value to allocate a number of protons to each signal or group of signals. If the molecular formula is not known estimate the ratio of the heights of the main integral steps to the nearest set of whole numbers.
2. Select from the more prominent features in the spectrum what you believe is likely to be the most readily interpreted signal or group of signals.
3. Use the chemical shift chart to assign a specific proton group (or set of reasonable alternatives) to the signal.
4. If the signal is a multiplet look for its coupled partner. Remember that both sets of signals are likely to lean towards each other and that each is split by the same coupling constant.
5. Assign a proton grouping to the coupled partner and then taking into account the number of lines in each signal and the size of the coupling constant* construct a molecular fragment which is consistent with all the data.
6. Repeat the process (2) to (5) on the other spectral bands.
7. Combine the various fragments so that they give a molecular structure which is consistent with the spectrum.

Now try the first problem.

---

* In all of the spectra in Part 6, 0.1 p.p.m. = 6 Hz.

PART 6: PROBLEMS

F267 Problem 1
In the first problem we are going to deduce the molecular structure of a pale yellow solid ($C_{10}H_{10}O$), from its n.m.r. spectrum (Fig. 6.1).

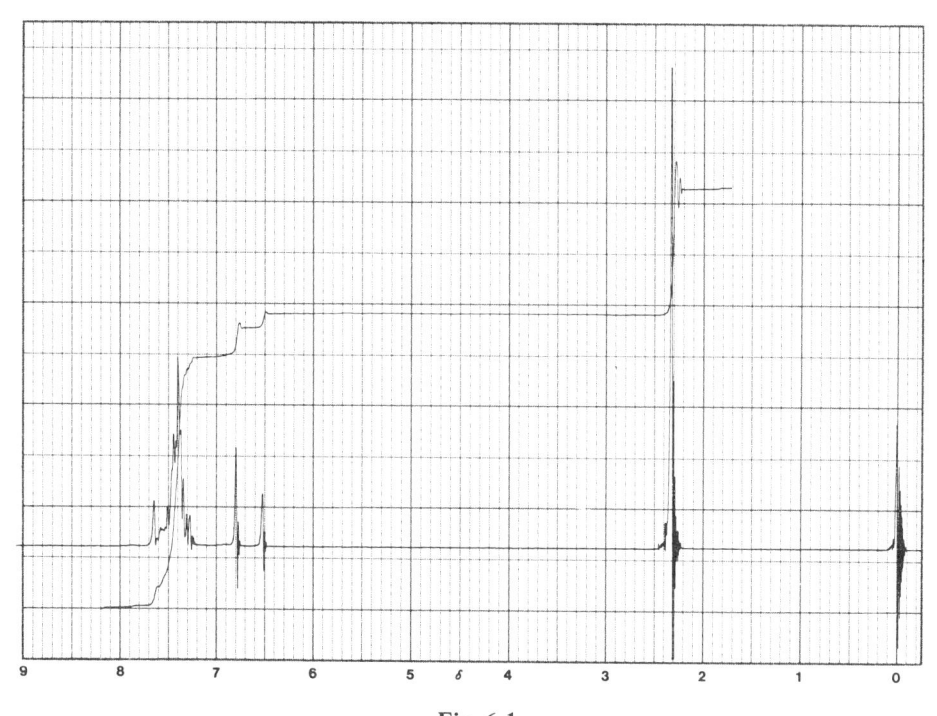

Fig. 6.1.

Estimate the number of protons per signal in the usual way and then begin the detailed analysis of the spectrum with the high field signal. Its shape and chemical shift can only be accounted for by one of two molecular fragments. What are they?

F268 The integration curve indicates that the two bands near 6.8 p.p.m. and 6.5 p.p.m. arise from one proton, that is, they constitute a doublet. Estimate the chemical shift of this proton as the weighted mean of the two signals and use the value together with the size of the coupling constant to deduce the nature of its environment.

    A267 7.2 to 7.7 p.p.m. = 6 protons; 6.5 to 6.8 p.p.m. = 1 proton; 2.3 p.p.m. = 3 protons.

    The high field signal is a sharp single band indicating that the protons are not coupled. The three protons probably belong to an isolated methyl group which, from the chemical shift value, is likely (in a compound lacking nitrogen or halogen) to be attached to a carbonyl group or an aromatic ring. The two molecular fragments are thus **1** or **2**.

$$CH_3-C\underset{O}{\overset{\nwarrow}{\diagup}} \quad \text{or} \quad CH_3-\text{C}_6H_5$$

        1   **2**

F269 The doublet implies that the 6.7 p.p.m. alkenic proton is coupled to one other proton. Moreover, the higher intensity of the low field component of the doublet indicates that its coupled partner is downfield. It must be amidst the complex multiplet between 7.2 and 7.7 p.p.m. Examine the latter carefully and pick out the coupled partner. Record its chemical shift.

    A268 Approximately 6.7 p.p.m. The chemical shift chart gives three possibilities, a proton attached to a benzene ring, a heterocycle or a conjugated alkene. Only the latter is compatible with the large coupling constant of the doublet.

F270 The low field doublet with its mean position at approximately 7.4 p.p.m. lies in the conjugated alkene or aromatic proton region of the spectrum. Support for the former comes from the J-value which is attributable to alkenic proton spin–spin coupling. Obtain this value and then select a molecular fragment from the following compounds **3**, **4** and **5** which accounts for the signals of both alkenic protons.

$$\underset{3}{\overset{H}{\diagdown}C=C\overset{H}{\diagup}} \quad , \quad \underset{4}{\overset{H}{\diagdown}C=C\overset{}{\underset{H}{\diagdown}}} \quad , \quad \underset{5}{\overset{}{\diagdown}C=C\overset{H}{\underset{H}{\diagdown}}}$$

    A269 Approximately 7.45 p.p.m. You should have been able to discern two lines at 7.65 and 7.35 p.p.m. which are practically the mirror image of the 6.7 p.p.m. doublet. The weighted mean of these approximates to 7.45 p.p.m.

F271 Five protons remain to be assigned. Formulate a possible molecular fragment containing these protons which accounts for the remainder of the 7.2 to 7.7 p.p.m. complex.

    A270 17 Hz; $\overset{H}{\diagdown}C=C\overset{}{\underset{H}{\diagdown}}$

                  4

F272 So far you have deduced that the molecule contains the fragments:

$$\text{(phenyl ring with H's)} \quad , \quad \overset{H}{\diagdown}C=C\overset{}{\underset{H}{\diagup}} \quad \text{and} \quad CH_3-C\overset{\nearrow O}{\diagdown} \quad \text{or} \quad CH_3-C_6H_5$$

      6   **4**   **7**

                                                      8

Assemble these to obtain a molecular structure which is consistent with the formula and n.m.r. spectrum.

A271 This signal occurs within the aromatic region. The fact that it corresponds to **five** protons suggests very strongly that the group responsible is phenyl (6).

$$\text{C}_6\text{H}_5- \quad \mathbf{6}$$

A272

$$\text{C}_6\text{H}_5-\underset{H}{\overset{H}{C}}=C-\overset{O}{\underset{}{\overset{\parallel}{C}}}-CH_3$$

9

F273 Problem 2

The second problem is to deduce the structure of a colourless solid ($C_{10}H_{13}NO$) from its spectrum (Fig. 6.2).

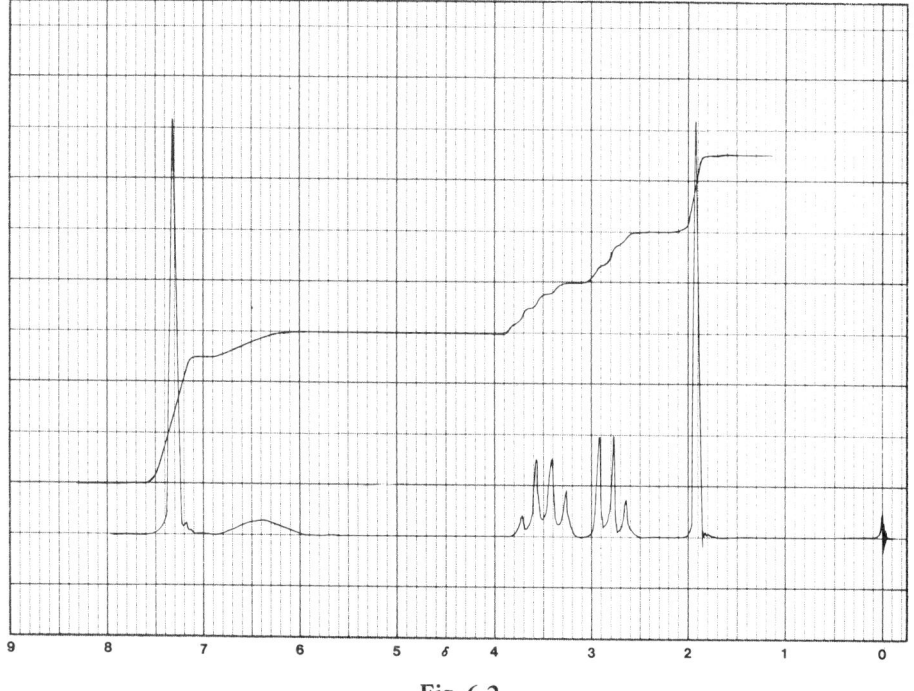

Fig. 6.2.

First use the spectrum to infer the number of sets of chemically different protons within the molecule and then estimate the actual number of protons in each set.

F274 Account for the two largest signals at 1.9 and 7.3 p.p.m.

> **A273** The spectrum shows five groups of lines, which correspond to five sets of protons in chemically distinct environments. The steps of the integral trace are in the ratio 5:1:2:2:3 and since the total number of protons is thirteen, these numbers represent the actual number of protons in each environment.

F275 Now turn your attention to the multiplets at 2.8 and 3.5 p.p.m. The position, integration and coupling of these two signals should enable you to deduce the structures of two different molecular fragments which are consistent with the molecular formula. Both fragments should incorporate the four protons directly responsible for the signals and at least three other atoms.

> **A274** The low field signal at 7.3 p.p.m. is characteristic of the spectra of aromatic compounds and in this case could be attributed to the absorption by the five protons of a monosubstituted benzene ring.
>
> The 1.9 p.p.m. singlet arises from three equivalent protons, most probably those of a methyl group, which from the chemical shift is likely to be attached to a carbonyl group or a double bond. The singlet implies there is no coupling to other protons and supports the assignment CH_3CO-. The alkene structure, on the other hand, has to be of the type **10** in order to be consistent with the apparent absence of vicinal or allylic (*trans*) coupling. Further reasoning excludes this on the following grounds. The only ways in which **10** and a benzene ring may be pieced together produce fragments such as **11** containing ten or more carbon atoms. As the molecule has in total only ten carbon atoms and three more signals remain to be accounted for then **10** must be eliminated and we can confine our attention to the CH_3CO- assignment.

F276 Only the signal at 6.4 p.p.m. remains. Account for this and in doing so justify the appearance of the absorption band.

> **A275** Each corresponds to absorption by two protons and hence probably arise from methylene groups. As the 2.8 p.p.m. signal is split into three lines it must be coupled equally to two other protons, which from the general appearance of the spectrum must be those of the other methylene group. The signal for this group (at 3.5 p.p.m.) appears as a quartet which suggests that it is not coupled to two protons but to three. Two of these belong to the 2.8 p.p.m. methylene group and the third, which appears to have, by chance, the same coupling as that between the two methylene groups, remains to be identified. We can therefore explain the two signals by the partial structure **12**.
>
> $-CH_2-CH_2-$
> 2.8   3.5 p.p.m.
> **12**

The high δ-value (3.5) of the methylene group means that it must be attached directly to an electronegative element such as halogen, oxygen or nitrogen. The molecular formula rules out halogen, and the previous assignment of the molecule's only oxygen atom to a carbonyl group eliminates ether oxygen. The partial structure **13** therefore results.

$$-CH_2-CH_2-N\!\!<$$
$$2.8 \quad 3.5 \text{ p.p.m.}$$
13

The methylene group at 2.8 p.p.m. may in a similar manner be placed next to nitrogen, a carbonyl group or a benzene ring. Having already assigned the molecule's only nitrogen atom we can reasonably expand the partial structure to either **14** or **15**.

Ph—CH₂—CH₂—N<   or   —CO—CH₂—CH₂—N<
2.8   3.5 p.p.m.
14   **15**

F277 Now that you have accounted for each signal assemble the molecular fragments to produce two different structures both of which are compatible with the spectrum and formula of the unknown.

A276 The broad band is characteristic of protons attached to nitrogen or oxygen. The latter is incompatible with earlier deductions. Support for the former comes from the number of bands in the 3.5 p.p.m. methylene signal. You should recall that when NH protons are not exchanging rapidly they do couple with neighbouring protons. In this case they split what otherwise would have been a triplet into a quartet. However unlike slowly exchanging OH protons the reciprocal coupling is not apparent because of the influence of the magnetic character of the nitrogen nucleus.

A277

Ph—CH₂CH₂NHCOCH₃   or   CH₃COCH₂CH₂NH—Ph
16   **17**

The spectrum is, in fact, that of compound **16**. They could be distinguished by infrared and ultraviolet spectroscopy. An experienced n.m.r. spectroscopist would also recognize that the singlet at 7.3 p.p.m. was compatible with an alkyl substituted benzene in which the chemical shift of the *ortho* protons is accidentally equivalent to that of the *meta* and *para*.

F278 Problem 3

The spectrum in Fig. 6.3 was obtained from a compound known to be either **18** or **19**. Its identity may be readily deduced from the integration and an examination of the low field signals (above 5 p.p.m.). Carry out this identification and indicate your reasoning.

18

19

Fig. 6.3.

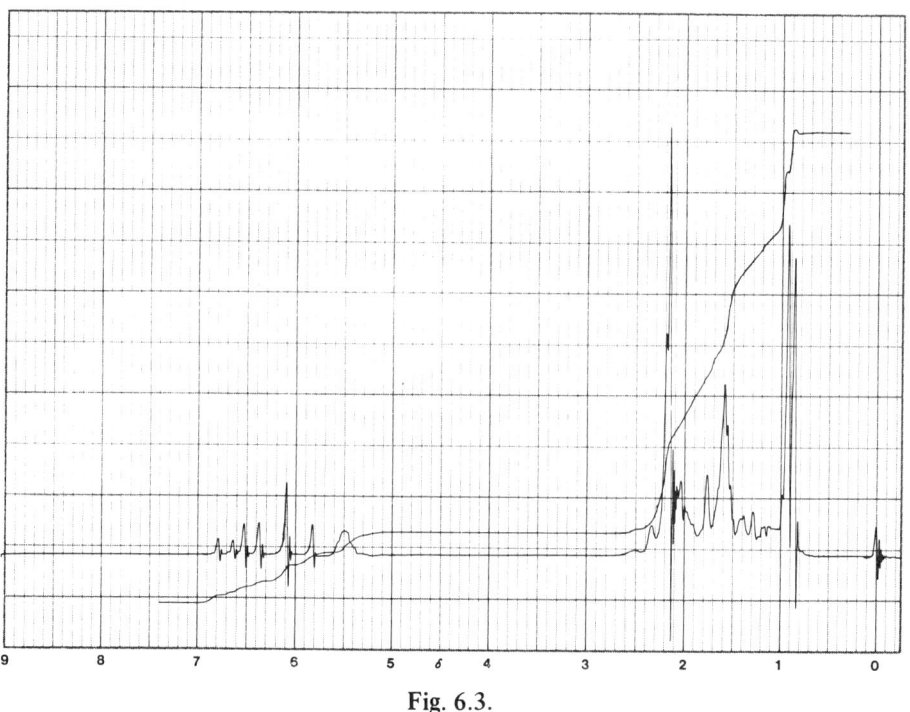

F279 Now that you have identified the compound it is instructive to pursue the problem further to ascertain how certain features of the spectrum have arisen. For example, how do you account for the appearance of the signals of the alkenic protons?

> **A278** **18**. The low field signals arise from the alkenic protons. Compound **19** has two such protons, which, being coupled together would be expected to give rise to a pair of doublets appearing as an *AB* quartet. The spectrum's seven line pattern at low field with its accompanying three proton integration value is clearly incompatible with this and hence structure **19** must be eliminated.

F280 Account for the pair of signals near 0.9 p.p.m.

A279 The six sharp signals arise from the protons H^a and H^b. Their position and appearance may be rationalized as follows. The electron density in the vicinity of H^a and H^b will be influenced by the electron withdrawing power of the carbonyl group, this effect being most pronounced at the β-carbon atom of the αβ unsaturated carbonyl system (see **20a** and **20b**).

$$\text{>C=C-C=O} \longleftrightarrow \text{>C}^+\text{-C=C-O}^-$$
$$\quad\quad\text{20a} \quad\quad\quad\quad\quad\quad \text{20b}$$

Consequently the proton attached to the β-carbon atom, that is H^b, will be de-shielded more than H^a and will resonate at a lower field position.

H^a is coupled to H^b and should therefore appear as a doublet; H^b is coupled to both H^a and H^c and hence should give rise to a doublet of doublets. The patterns are illustrated by the branching diagram shown.

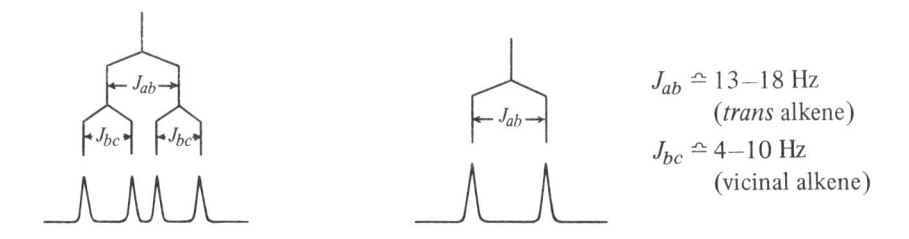

$J_{ab} \simeq 13-18$ Hz (*trans* alkene)

$J_{bc} \simeq 4-10$ Hz (vicinal alkene)

The observed variation of intensity from the predicted values results from the relatively small chemical shift differences between H^a and H^b leading to a departure from a simple first-order spectrum.

The seventh signal, the rather diffuse band at 5.5 p.p.m. arises from the isolated alkenic proton H^d which is not only coupled to the two adjacent methylene protons but is also weakly coupled to allylic protons H^c and H^e.

A280 The pair of signals arise from the components of the *gem*-dimethyl groups CH_3^f and CH_3^g. Although at first glance they may appear equivalent they are, in fact, magnetically non-equivalent owing to their positions on different sides of an unsymmetrically substituted ring system (**21**). They are, in fact, next to the chiral centre A.

F281 Problem 4

The spectrum in Fig. 6.4 is that of a nitrocompound $C_3H_6ClNO_2$. Deduce its structure and account for the complexity of the signal near 2.3 p.p.m.

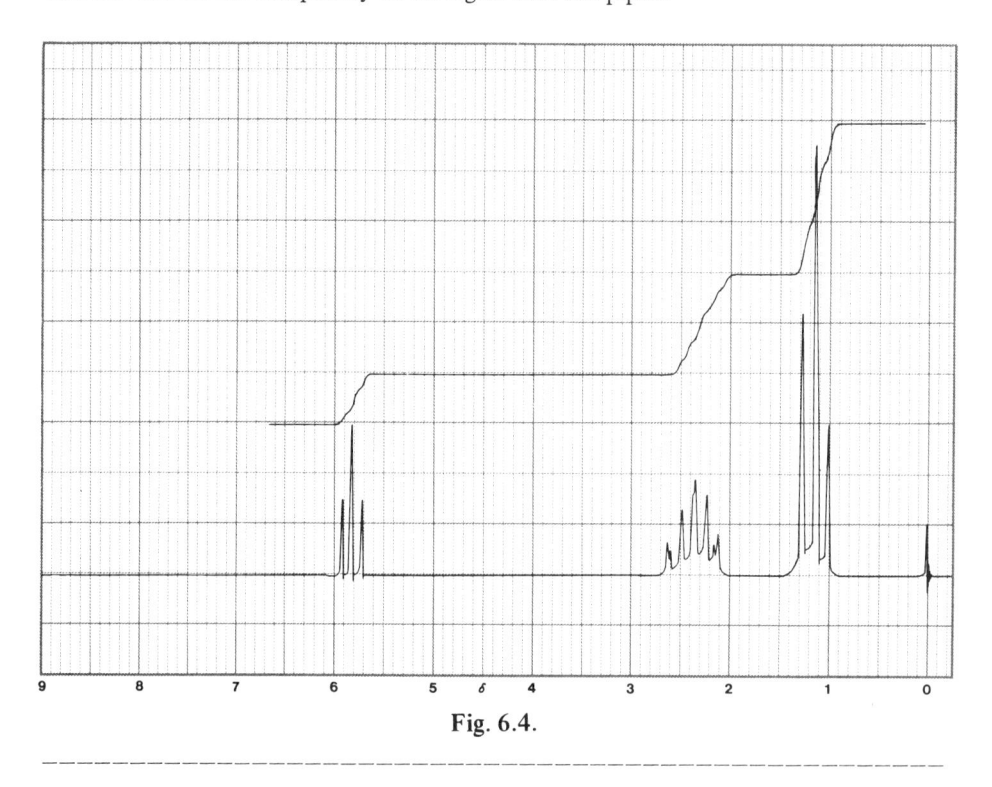

Fig. 6.4.

A281 The integration curve indicates that the compound's six protons are distributed between three broadly magnetically different environments in the ratio 1:2:3. In the absence of C=C (precluded by the molecular formula) the chemical shift of the lone proton signal at 5.8 p.p.m. is particularly low. This suggests that it is deshielded by the inductive effect of both the nitro group and the halogen

$$\begin{array}{ccc} NO_2 & NO_2 & Cl \\ | & | & | \\ H-C & H-C-CH_2- & CH_3CH_2CH \\ | & | & | \\ Cl & Cl & NO_2 \\ \mathbf{22} & \mathbf{23} & \mathbf{24} \end{array}$$

atom. The molecular fragment **22** can therefore be written and then a methylene group added (*viz.* **23**) to account for the triple line signal. Only one C and three H's remain to be accounted for. These must belong to a methyl group which being attached to the $-CH_2-$ gives rise to the triplet at 1.1 p.p.m. The compound is therefore 1-chloro-1-nitropropane (**24**). The signal near 2.3 p.p.m. arises from the methylene protons. These are flanked by the methyl and methine protons, a situation which gives rise to either a doublet of quartets, or a quartet of doublets or more usually, on account of fortuitously equal coupling constants, a pentuplet.

The actual signal is more complicated than any of these and to account for this you should have noticed that the methylene protons are next to a chiral centre and consequently are magnetically non-equivalent. As such they can be expected to have slightly different chemical shifts and couple not only to the protons on the adjacent atoms but also to each other.

F282 Problem 5

Deduce the structure of the compound $C_6H_{10}O_2$ whose n.m.r. spectrum is shown in Fig. 6.5.

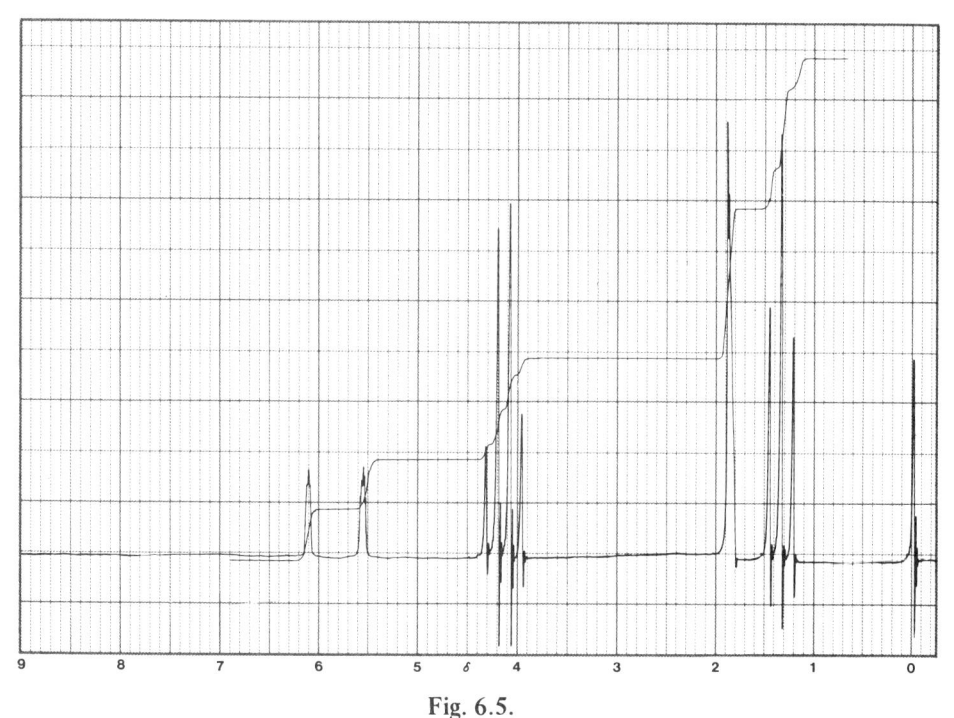

Fig. 6.5.

A282 There appear to be five groups of signals in the spectrum with intensities in the ratio 1:1:2:3:3. As the total number of protons in the molecule is ten, this ratio represents the actual numbers of protons responsible for each signal.

Perhaps the most striking feature of the spectrum is the triplet and quartet signals near 1.3 p.p.m. and 4.2 p.p.m. each with the same band separation of 7 Hz. You should have concluded that these were the components of an A_2X_3 spin–spin multiplet system which, on the basis of the quartet's chemical shift value, should be attributed to an ethoxyl group ($-OCH_2CH_3$). The two low field signals at 6.1 p.p.m. and 5.6 p.p.m. each arise from a single proton. Both show signs of weak coupling such as that expected from long range coupling or from geminal coupling of alkene protons. The molecular formula points to the latter and hence we can write a second molecular fragment (25).

$$\begin{array}{c} H \\ H \end{array} C=C \begin{array}{c} \\ \\ \end{array}$$

25

The remaining signal at 2.0 p.p.m. corresponds to three protons which from the δ-value may be assigned to a methyl group attached to either >C=O or >C=C<. Although the molecular formula permits either of these, the former may be ruled out because of the fine splitting which is discernible on close inspection of the signal. This splitting implies spin–spin coupling which is not normally shown by the protons of methyl ketones. The methyl group must therefore be attached to a double bond and the appearance of the signal may be attributed to the relatively weak coupling of *trans* allylic protons. The third fragment may thus be drawn as shown in **26**. With only one carbon and one oxygen atom remaining, only structure **27** can be drawn consistent with all the data.

$$\underset{H}{\overset{}{>}}C=C\underset{}{\overset{CH_3}{<}}$$

26

$$\underset{H}{\overset{H}{>}}C=C\underset{\underset{O}{\overset{\|}{C}-O-CH_2CH_3}}{\overset{CH_3}{<}}$$

27

Answers to frames 283 to 285 will be found on p. 176. Try all three problems before consulting them.

F283 Problem 6
A dihaloalkane C_3H_6BrCl gave the n.m.r. spectrum shown in Fig. 6.6. What is its structure?

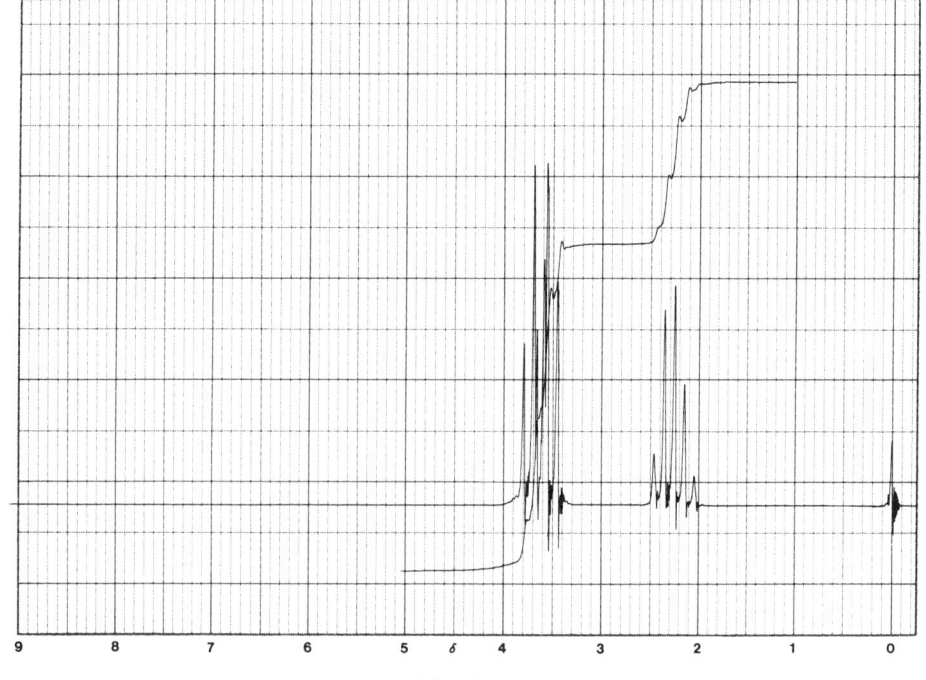

Fig. 6.6.

F284 Problem 7

The spectrum of an aromatic dinitrocompound $C_7H_6N_2O_5$ is shown in Fig. 6.7. Deduce its structure.

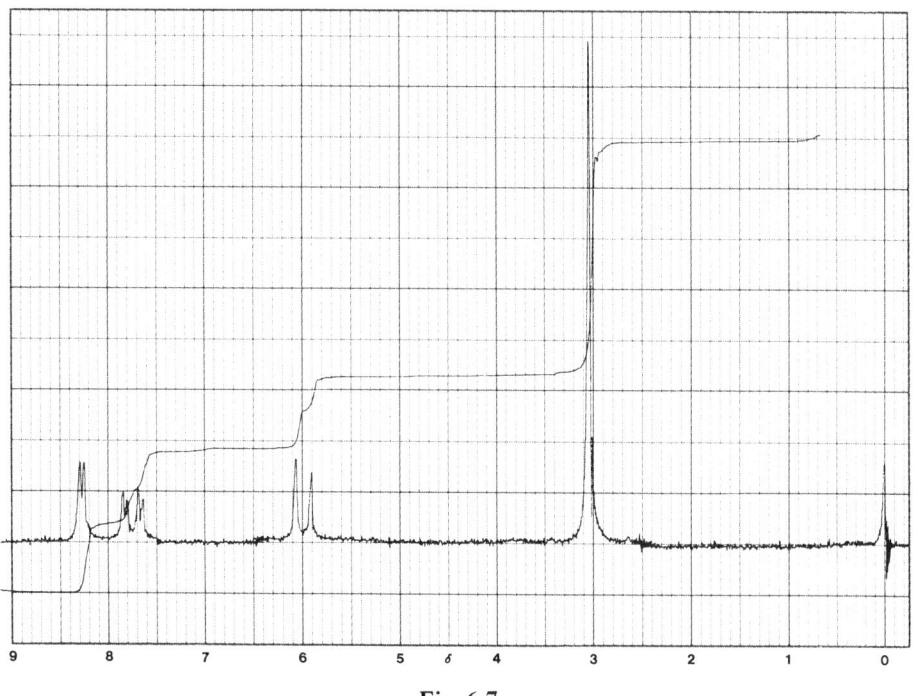

Fig. 6.7.

F285 Problem 8

Deduce the structure of the carboxylic acid $C_{12}H_{14}O_2$ whose n.m.r. spectrum is shown in Fig. 6.8. The signal of the COOH proton is not shown as it lies off the chart between 10 and 13 p.p.m.

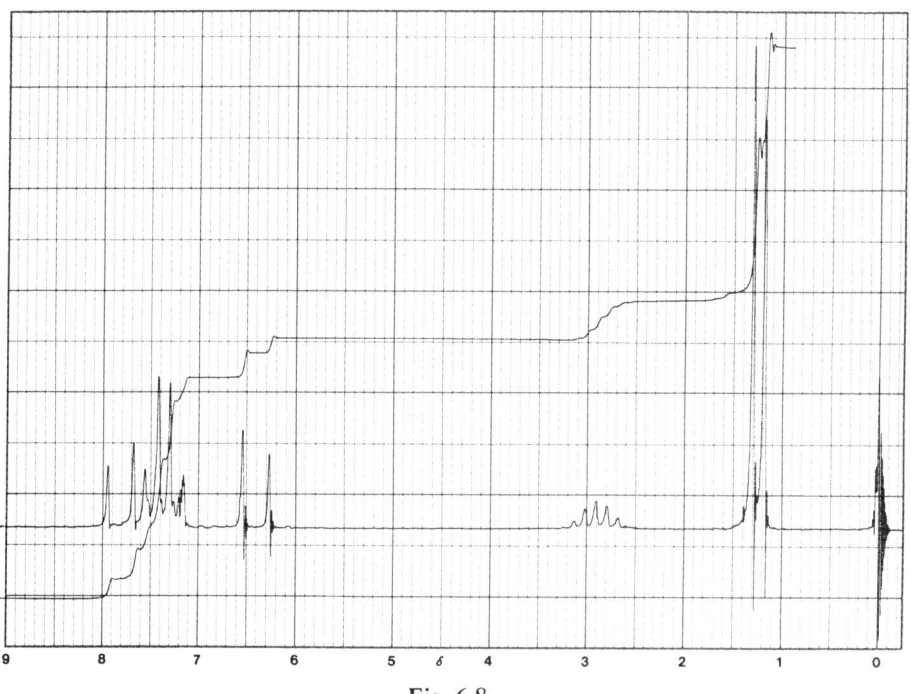

Fig. 6.8.

Further Reading

R. M. Silverstein and G. C. Bassler, *Spectrometric Identification of Organic Compounds*, John Wiley and Sons, London, 1968, Chapter 6.

J. R. Dyer, *Applications of Absorption Spectroscopy of Organic Compounds*, Prentice Hall, New Jersey, 1965, Chapter 4.15.

A. Ault, *Problems in Organic Structure Determination*, Part II, McGraw-Hill, 1967.

D. H. Williams and I. Fleming, *Spectroscopic Problems in Organic Chemistry*, McGraw-Hill, London, 1967.

B. M. Trost, *Problems in Spectroscopy*, Benjamin, New York, 1967.

T. Cairns, *Spectroscopic Problems in Organic Chemistry*, Heyden and Son Ltd., London, 1964.

A. J. Baker, T. Cairns, G. Eglinton and F. J. Preston, *More Spectroscopic Problems in Organic Chemistry*, Heyden and Son Ltd., London, 1967.

M. B. Winstead, *Organic Chemistry Structural Problems*, Sadtler Research Laboratories, Philadelphia, 1968.

D. J. Pasto and C. R. Johnson, *Organic Structure Determination*, Prentice Hall, New Jersey, 1969, Chapter 5.

F. Scheinmann, *An Introduction to Spectroscopic Methods for the Identification of Organic Compounds*, Volume 1, Pergamon Press, Oxford, 1970.

Answers to Multiple Choice Tests

Part 1
1. a
2. a
3. b
4. c
5. c
6. b
7. c
8. b
9. a
10. b

Part 2
1. b
2. a
3. a
4. b
5. b
6. b
7. c
8. c
9. b
10. b

Part 3
1. a
2. c
3. c
4. b
5. a
6. **3**
7. b
8. 8
9. c
10. a

Part 4
1. c
2. c
3. **1**
4. a
5. a
6. c
7. c
8. a
9. c
10. b

Part 5
1. **1**
2. c
3. a
4. b
5. 7
6. c
7. c
8. a
9. b
10. c

Selected Answers to Questions

Part 1
2. (a) 6 hydrogen atoms; (b) 4 hydrogen and 2 nitrogen atoms; (c) 1 fluorine and 1 hydrogen atom.

Part 2
1. $-OCH_3$ and $-COCH_3$, or $-CH_3$ and $-COOCH_3$.

2. $CH_2=C-C\equiv CH$
    $\quad\quad\quad |$
    $\quad\quad\ CH_3$

3. (a) 6.2; (b) 4.5; (c) 1.9; (d) 1.8 and (e) 3.8 p.p.m.

4. Compound **64**.

5. 75% of the benzaldehyde had been reduced to benzyl alcohol.

Part 3
2. (a) **50**; (b) **52**; (c) **48**; (d) **49**; (e) **51**.

3. [thiazole-2-amine structure] [5-amino-isoxazole structure] [3-amino-isothiazole structure]

Part 5
4. [1-methoxy-1-isopropyl-benzene type structure: Ph-C(OMe)(H)-CH(CH_3)_2]

Answers to Problems 6, 7 and 8

A283	$ClCH_2CH_2CH_2Br$

A284	[2,4-dinitroanisole structure] **28**

A285	[4-isopropyl-cinnamic acid structure: (CH_3)_2CH-C_6H_4-CH=CH-COOH, trans] **29**

List of Compounds

Compounds discussed in the programme include the following:

Acetone, 36
N-Acetyl-2-phenylethylamine, 165
18-Annulene, 52

Barbaralone, 140
4-Bromobutyronitrile, 113

Chloroacetaldehyde, 76
1-Chloro-3-bromopropane, 176
trans-1-Chloropropene, 111
p-Chlorostryene, 96
Cyclohexane, 138

Decamethylenebenzene, 52
1,2-Dibromo-1-phenylethane, 150
Diethyl-N-acetyl-2-aminomalonate, 88
Diethyl acetyl succinate, 145
9,10-Dihydroanthracene, 47
Dihydrobullvalene, 141
2,3-Dihydropyran, 120
Dimethylformamide, 137
2,4-Dinitroanisole, 176

Ethanol, 21, 70, 81, 142
Ethyl acetate, 73
Ethyl chloroacetate, 86

Ethylene chlorohydrin, 143
Ethyl methacrylate, 70
Ethyl vinyl ether, 93, 116

α and β-Ionone, 166
p-Isopropylcinnamic acid, 176

Methanol, 139
Methyl acetate, 38
2-Methylbutan-1-ol, 133
4-Methylbutyrolactone, 90
N-Methylurethane, 84

1-Nitropropane, 115

Pentan-1-ol, 125
Phenylalanine, 152

Semibullvalene, 141
p-Styryl methyl ketone, 163

Tetramethylsilane, 35, 58
2,2,4-Trimethylpentane-1,3-diol, 146
Toluene, 44
Tris-(2,2,6,6-tetramethylheptane-3,5-dionato) europium, 125

Subject Index

AB quartet (calculation), 108
Absorption, 11, 15
Accidental equivalence, 145, 153
Aldehydes, 50, 57
Alkenes, 49, 56
Alkynes, 54, 57
Alphabetical notation, 109, 126
Anisotropy, 54
Applied magnetic field, 7
Aromaticity, 51
Aromatic ring current, 51, 57
Aryl protons (chemical equivalence), 45
Asymmetric carbon, 143
Atomic number, 7

Benzene—
    ring currents in, 51, 57
    solvent effects, 55
Branching method, 76, 97

Carbon-13, 8
Chemical equivalence, 27, 28, 45, 55, 110, 126, 142, 153
Chemical shift, 21, 34, 58
    calculation for AB quartet, 108
    empirical values, 40, 42
    NH and OH protons, 42, 58
    reagents, 123
Chirality, 143, 153
Computed spectra, 122
Coupling *(see spin–spin coupling)*
Coupling constant, 72, 97
    empirical values, 90

Delta (δ)-scale, 37, 58
Deshielding, 49, 56
Deuterated solvents, 13
Diamagnetic anisotropy, 55
Diamagnetic shielding, 26, 55
    (see also shielding)
Distortion of signals, 107
Double resonance, 114

Empirical correlations—
    chemical shifts, 40, 42, 58
    coupling constants, 90
Energy levels, 9, 14

Exchange—
    deuterium, 120
    protons, 42, 58, 81, 98, 139

First-order spectra, 111

Geminal magnetic non-equivalence, 142, 153

High field strength, 215
Hindered rotation, 137
Hydrogen bonding, 42, 58, 81, 98

Induced magnetic field, 25
    multiple bonds, 49
    single bonds, 25
Inductive effect, 27, 56
Integration, 22, 30, 57
Interpretation of spectra, 160

Magnetic equivalence, 27, 28, 55, 110, 126, 142, 153
Magnetic moment, 9
Mass number, 7

NH protons—
    chemical shift, 42, 58
    coupling, 81, 98
Nuclear magnetic moment, 9
Nuclear magnetic resonance, 7, 14
    spectrometer, 10
    time scale, 137
Nuclear spin, 7

OH protons—
    chemical shift, 42, 58
    coupling, 81, 98

Paramagnetic shift reagents, 123
Parts per million (p.p.m.), 37

Relaxation, 11
Ring current, 51, 57
Ring inversion, 138

Sample preparation, 13, 15
Saturation, 11, 14
Shielding, 25, 55

Shielding *(contd.)*
    multiple bonds, 48, 56
    single bonds, 25, 55
Shift reagents, 123
Solid state spectra, 12
Solvents, 13, 15
Spin, 7
Spin quantum number, 8, 14
Spin–spin coupling, 21
    alphabetical notation, 109, 126
    branching method, 76
    decoupling, 114

Spin–spin coupling *(contd.)*
    NH and OH protons, 81, 98
    origin, 70, 97

Tau (τ)-scale, 38, 58
Temperature dependent spectra, 137, 153
Tetramethylsilane (TMS), 35, 58

Units of chemical shift, 34, 58

Valence isomerization, 140
Variable temperature spectra, 137, 153